T0212976

SpringerBriefs in Applied Sciences and Technology

Computational Intelligence

Series editor

Janusz Kacprzyk, Polish Academy of Sciences, Systems Research Institute, Warsaw, Poland

The series "Studies in Computational Intelligence" (SCI) publishes new developments and advances in the various areas of computational intelligence—quickly and with a high quality. The intent is to cover the theory, applications, and design methods of computational intelligence, as embedded in the fields of engineering, computer science, physics and life sciences, as well as the methodologies behind them. The series contains monographs, lecture notes and edited volumes in computational intelligence spanning the areas of neural networks, connectionist systems, genetic algorithms, evolutionary computation, artificial intelligence, cellular automata, self-organizing systems, soft computing, fuzzy systems, and hybrid intelligent systems. Of particular value to both the contributors and the readership are the short publication timeframe and the world-wide distribution, which enable both wide and rapid dissemination of research output.

More information about this series at http://www.springer.com/series/10618

Frumen Olivas · Fevrier Valdez
Oscar Castillo · Patricia Melin

Dynamic Parameter Adaptation for Meta-Heuristic Optimization Algorithms Through Type-2 Fuzzy Logic

 Springer

Frumen Olivas
Division of Graduate Studies
Tijuana Institute of Technology
Tijuana, Baja California
Mexico

Oscar Castillo
Division of Graduate Studies
Tijuana Institute of Technology
Tijuana, Baja California
Mexico

Fevrier Valdez
Division of Graduate Studies
Tijuana Institute of Technology
Tijuana, Baja California
Mexico

Patricia Melin
Division of Graduate Studies
Tijuana Institute of Technology
Tijuana, Baja California
Mexico

ISSN 2191-530X   ISSN 2191-5318   (electronic)
SpringerBriefs in Applied Sciences and Technology
ISSN 2520-8551   ISSN 2520-856X   (electronic)
SpringerBriefs in Computational Intelligence
ISBN 978-3-319-70850-8   ISBN 978-3-319-70851-5   (eBook)
https://doi.org/10.1007/978-3-319-70851-5

Library of Congress Control Number: 2018930140

© The Author(s) 2018
This work is subject to copyright. All rights are reserved by the Publisher, whether the whole or part of the material is concerned, specifically the rights of translation, reprinting, reuse of illustrations, recitation, broadcasting, reproduction on microfilms or in any other physical way, and transmission or information storage and retrieval, electronic adaptation, computer software, or by similar or dissimilar methodology now known or hereafter developed.
The use of general descriptive names, registered names, trademarks, service marks, etc. in this publication does not imply, even in the absence of a specific statement, that such names are exempt from the relevant protective laws and regulations and therefore free for general use.
The publisher, the authors and the editors are safe to assume that the advice and information in this book are believed to be true and accurate at the date of publication. Neither the publisher nor the authors or the editors give a warranty, express or implied, with respect to the material contained herein or for any errors or omissions that may have been made. The publisher remains neutral with regard to jurisdictional claims in published maps and institutional affiliations.

Printed on acid-free paper

This Springer imprint is published by Springer Nature
The registered company is Springer International Publishing AG
The registered company address is: Gewerbestrasse 11, 6330 Cham, Switzerland

Preface

Other works that deal, partially, the improvement to the original methods of

In this book, a methodology for parameter adaptation in meta-heuristic optimization methods is proposed. This methodology is based on using metrics about the population of the meta-heuristic methods to decide through a fuzzy inference system the best parameter values that were carefully selected to be adjusted. With this modification of parameters, we want to find a better model of the behavior of the optimization method, because with the modification of parameters, these will affect directly the way in which the global or local search is performed.

Three different optimization methods were used to verify the improvement of the proposed methodology. In this case, the optimization methods are particle swarm optimization (PSO), ant colony optimization (ACO), and gravitational search algorithm (GSA), where some parameters are selected to be dynamically adjusted, and these parameters have the most impact in the behavior of each optimization method.

Simulation results show that the proposed methodology helps each optimization method in obtaining better results than the results obtained by the original method without parameter adjustment.

In Chap. 1, a brief introduction of the benefit of using the proposed methodology for parameter adaptation is presented, and an overview of the problem that we try to solve with the methodology.

In Chap. 2, a background and general concepts, needed to understand this research, like the fuzzy inference system, and all the optimization algorithms used in this book are presented.

Chapter 3 presents a description of all the problems used to test all the optimization methods with parameter adaptation using the proposed methodology.

In Chap. 4, the methodology for parameter adaptation and a detailed explanation of how the parameter adaptation was performed in all the optimization methods are presented.

Chapter 5 presents simulations and results of the optimization methods with parameter adaptation using the proposed methodology in all the problems described in Chap. 3, where results with the original optimization methods and even some other works that also perform an improvement in the original algorithm are presented.

Chapter 6 presents a statistical comparison of results between the original optimization methods and the optimization methods with parameter adaptation using the proposed methodology.

Finally, in Chap. 7, the conclusions and future work are presented, followed by the appendix with the code.

We end this preface of the book by giving thanks to all the people who have helped or encouraged us during the writing of this book. First of all, we would like to thank our colleague and friend Prof. Janusz Kacprzyk for always supporting our work and for motivating us to write our research work. We would also like to thank our colleagues working in Soft Computing, which are too many to mention each by their name. Of course, we need to thank our supporting agencies, CONACYT and DGEST, in our country for their help during this project. We have to thank our institution, Tijuana Institute of Technology, for always supporting our projects. Finally, we want to thank our families for their continuous support during the time that we spend on this project.

Tijuana, Mexico   Frumen Olivas
                                                                             Fevrier Valdez
                                                                             Oscar Castillo
                                                                             Patricia Melin

Contents

Chapter 1
Introduction

Optimization is a process where an algorithm creates several variations of solutions to a problem and in an intelligent way can improve these solutions iteratively, using meta-heuristics based on nature like evolution of species, the way in which ants find their food from their nest, or the way in which flock of birds fly together, more known as collective intelligence.

Bio-inspired optimization algorithms can be applied to a wide variety of problems but lack the ability to change dynamically their parameters to self-adapt into different problems; in this book, we present a methodology for parameter adaptation using fuzzy logic.

The proposed methodology can be applied to any optimization method that meets the specification criteria from the general procedure of the proposed methodology. Also, another metrics can be applied for more specific optimization methods.

There is a huge problem in setting the correct parameters for an optimization method when it is applied to a new problem; we can just use the recommended parameters for most common problems but these are not always the best parameters and depending on the problem can be the worst; this is why there are methodologies for parameter adaptation using mathematical methods, random parameters, and lastly fuzzy logic.

The use of fuzzy logic brings the tools needed to model a complex problem with easy to understand concepts like membership functions, for small, medium, or high values of a variable from the problem and use if-then rules to create complex knowledge about the problem in an easy way.

With a dynamic parameter adaptation, an optimization algorithm can obtain better quality results when compared with its original counterpart, also it can improve the diversity of solutions and the convergence of the algorithm, just by controlling one or more parameters dynamically over the iteration process.

© The Author(s) 2018
F. Olivas et al., *Dynamic Parameter Adaptation for Meta-Heuristic Optimization Algorithms Through Type-2 Fuzzy Logic*, SpringerBriefs in Computational Intelligence, https://doi.org/10.1007/978-3-319-70851-5_1

Chapter 2
Theory and Background

In this chapter, basic concepts of the main algorithms and theory used in this book are presented.

2.1 Fuzzy Logic

Fuzzy logic is an extension of traditional logic where there are only two possibilities true or false; meanwhile, in fuzzy logic this can be a range of values or labels. Fuzzy logic helps to model linguistic information using linguistic labels stipulated by membership functions.

2.1.1 Type-1 Fuzzy Inference Systems

Fuzzy logic was proposed in 1965 by Zadeh in [1–3], in an attempt to extend the traditional logic where there are only two options true or false, but in the case of fuzzy logic it can be more options like more or less false, almost true, etc.; mathematically, it is an extension of the traditional set where an element belongs or not to a given set. In fuzzy logic, an element can belong to a set just a little, i.e., a set named old_people in traditional set needs boundaries like all the people with more than 70 years belongs to that set, but a person with 69 years can not belong to this set. On the other hand, in fuzzy logic, the boundaries of a set are more "soft" which means that a person with 69 years can belong to the set old_people with 0.95 membership (from a membership range of 0–1). These characteristics of fuzzy logic give the tools needed to create models more easily, because they are more intuitive.

© The Author(s) 2018
F. Olivas et al., *Dynamic Parameter Adaptation for Meta-Heuristic Optimization Algorithms Through Type-2 Fuzzy Logic*, SpringerBriefs in Computational Intelligence, https://doi.org/10.1007/978-3-319-70851-5_2

2.1.2   Interval Type-2 Fuzzy Inference Systems

Interval type-2 fuzzy logic, also proposed by Zadeh and developed by Liang and Mendel in [4], is an extension of fuzzy logic where the fuzzy system can handle in a better way the uncertainty with a new type of membership functions, because they have a level of FOU (footprint of uncertainty); it is like having two type-1 membership functions, so the membership of a value in a fuzzy set now can be an interval of values and not only one value. This helps in the model of complex problems because now there is no need to set a specific membership for a value to a fuzzy set, and the membership can change between the upper and lower values of the interval [5].

In general with the use of interval type-2 fuzzy logic, we can design a better fuzzy system because we can build a better model of the uncertainty in real-world problems, and this is done using a footprint of uncertainty in the membership functions, which can now give a range of values of membership instead of a single number. Figure 2.1 illustrates the difference between a type-1 fuzzy set and an interval type-2 fuzzy set, with a particular value and its representation of degree of membership in each case.

In Fig. 2.1 for the same crisp value of 0.6, we have a 0.5 membership value in a type-1 triangular membership function representation. However, in an interval type-2 representation, the same crisp value 0.6 has a range of possible values of membership in the range from 0.24 to 0.72, so in the case that we do not know the exact membership degree of a crisp value in a fuzzy set. We can use interval type-2 fuzzy logic and set a range in which the value of membership can be found.

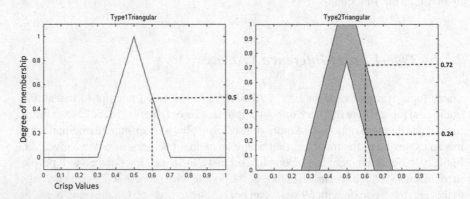

Fig. 2.1 Difference between type-1 and type-2 fuzzy sets (values are approximations for illustrative purposes)

2.2 Particle Swarm Optimization

Kennedy and Eberhart in 1995 [6, 7] introduce PSO as a bio-inspired method, which maintains a population or swarm of particles, where each particle represents a solution of the problem. Every particle "fly" is in the search space, using Eq. (2.1) to update the position of the particle and Eq. (2.2) to update the velocity of the particle [8]. The velocity equation is affected by a coefficient component and a social component, and at the same time these components are affected by coefficient factor c_1 and social factor c_2, respectively. These factors are very important because they affect directly the velocity of all the particles; so controlling these parameters, we can control the exploration and exploitation of PSO in the search space [9].

$$x_i (t + 1) = x_i (t) + v_i (t + 1) \qquad (2.1)$$

$$v_{ij} (t + 1) = C \left[v_{ij} (t) + c_1 r_1 (t) \left[y_{ij} (t) - x_{ij} (t) \right] + c_2 r_2 (t) \left[\hat{y}_j (t) - x_{ij} (t) \right] \right], \qquad (2.2)$$

where in Eq. (2.1), x_i is the position of the particle i and v_i is the velocity of the particle i, and in Eq. (2.2), C is the constriction factor, v_{ij} is the velocity of the particle i in the dimension j, c_1 is the cognitive factor, r_1 is a random value in the range of [0, 1], c_2 is the social factor, r_2 is a random value in the range of [0, 1], y_{ij} is the dimension j of the best position found so far by the particle i, x_{ij} is the dimension j of the current position of the particle i, and $\hat{y}_j$ is the dimension j of the best position of the swarm found so far.

Since PSO was introduced, several variations have been developed as variants of PSO, for example, PSO with coefficient component only, PSO with social component only, PSO with constriction factor, PSO with inertia weight, and some others. In this research, we are using PSO with constriction factor based on experiments that we perform before; we conclude that this variant can obtain better results than the others, and we want to improve its results with the proposed methodology.

The constriction factor helps PSO to improve the convergence of the algorithm, by controlling the velocity of the particles, and this parameter affects the new velocity not only the previous like the inertia weight variant of PSO. So in PSO with constriction factor, this parameter became the most important because with it we can control the new velocity of the particles and with this control the exploration and exploitation abilities of PSO.

2.3 Ant Colony Optimization

Dorigo in 1992 [10] introduced ACO and this method is based on the behavior of ant colonies to find food using the collective intelligence of the ants, with the use of pheromone trail in real ants; the population in ACO are known as artificial ants and each ant represents a solution of the problem.

ACO uses Eq. (2.3) to create the tours of the ants based on the probabilities of an ant k to select the next node from city i to city j. Equation (2.4) is used to calculate the evaporation of the pheromone trail. Equation (2.5) updates the pheromone trail based on the number of ants that passes through an arc, while Eq. (2.6) allows only the best ants to deposit pheromone in the arcs of the tour constructed.

$$P_{ij}^k = \frac{[\tau_{ij}]^\alpha [\eta_{ij}]^\beta}{\sum l \in N_i^k [\tau_{il}]^\alpha [\eta_{il}]^\beta} \quad \text{if } j \in N_i^k \tag{2.3}$$

$$\tau_{ij} \leftarrow (1-\rho)\,\tau_{ij}, \quad \forall\,(i,\,j) \in L \tag{2.4}$$

$$\tau_{ij} \leftarrow \tau_{ij} + \sum_{k=1}^{n} \Delta\tau_{ij}^k, \quad \forall\,(i,\,j) \in L \tag{2.5}$$

$$\Delta\tau_{ij}^k = \begin{cases} \frac{1}{C^k}, & \text{if the arc } (i,\,j) \text{ belongs to } T^k \\ 0, & \text{otherwise} \end{cases} \tag{2.6}$$

where in Eq. (2.3), τ represents the level of pheromone in an arc, α is the importance of the pheromone, η is the heuristic information weighted by the distance of the arc, β is the importance of the heuristic information, and N is a set of cities that the ant k has not visited yet. In Eq. (2.4), ρ represents the level of evaporation of the pheromone trails and L indicates all the arcs in the graph. In Eq. (2.5), Δ is a weight of the pheromone deposited in the arcs and is defined in Eq. (2.6). In Eq. (2.6), C is the total distance of the tour and T is a set of the cities visited by the ant k.

Since ACO was introduced, some improvements have been developed as variants of ACO, for example, max-min ant system, ant system rank-based, and ant colony system. In this research, we are using the variant ant system rank-based, and this variant was selected based on several experiments with all the variants.

The ACO variant used in this research has the advantage over the others, because it uses a mechanism of elitism where it rearranges the ants by their fitness, and the best ants can deposit more pheromone than the others to reinforce the best routes.

2.4   Gravitational Search Algorithm

The gravitational search algorithm (GSA) was proposed by Rashedi et al. in [11], which is a population-based algorithm that works with the fundamental principles of gravity and motion. In GSA, individuals are known as agents, so every agent is a solution to the problem in which GSA is implemented, the performance or fitness of the agents corresponds to their masses, the agent with the bigger mass corresponds to the best agent, every agent causes a gravity force that attracts all of the other agents, and this causes a movement of every agent toward an agent with bigger mass.

In GSA, the agents cooperate each other through the gravitational force; using Eq. (2.7), heavier agents are the best solutions that move slowly but attract more lighter agents. GSA uses the fitness function to determine its masses. The agents navigate through the search space by properly adjusting their gravitational force, inertia mass, and using the acceleration to move toward a better position over the iterations. The mass of the agents represents an optimum solution found in the search space [11].

$$F_{ij}^d(t) = G(t) \frac{M_{pi}(t) * M_{aj}(t)}{R_{ij}(t) + \varepsilon} \left(x_j^d(t) - x_i^d(t) \right), \tag{2.7}$$

where M_{aj} is the active mass of agent j, M_{pi} is the passive mass of agent i, G is the gravitational constant, ε is a small value used to avoid division by zero, and R_{ij} is the Euclidian distance between the agents.

GSA moves their agents toward the best agents using Eq. (2.8); this equation corresponds to the acceleration of an agent toward an agent with a bigger mass that used its gravitational force to generate an acceleration using Eq. (2.8).

$$a_i^d(t) = \frac{F_i^d(t)}{M_{ii}(t)}, \tag{2.8}$$

where M_{ii} is the inertial mass of agent i. This acceleration now is added to its current velocity of agent i. And now to determine the new position of the agent i, its new velocity is added to their current position.

In nature, the gravitational constant G is fixed but in GSA this can be variable and is calculated using Eq. (2.9). This gravitational constant is used by GSA to control the gravity force that agents can act on other agents. Therefore, controlling the gravitational constant, we can control the level of gravitational force.

$$G(t) = G_0 e^{-\alpha * t / T}, \tag{2.9}$$

where G_0 is the actual value of the gravitational constant G, α is a parameter user-defined which allows changing G over the iterations, t is the current iteration, and T is the total iterations.

The elitism in GSA only allows the best agents to apply their gravitational force to the remaining agents. This kind of elitism brings to GSA a way to have equilibrium in exploration and exploitation of the search space.

The parameter that controls the number of agents allowed to apply their force is Kbest, which means that only the Kbest agents will attract all of the other agents. In GSA, the way in which this parameter works is at initial iterations; most agents apply their force, and over the iterations that number decreases linearly and at final iterations, only a small percentage of agents will apply their force to the remaining agents [11]. So after the gravity force of all agents is calculated and before the calculation of the acceleration, the gravity forces of the agents change with Eq. (2.10).

$$F_i^d(t) = \sum_{j \in K_{best}, j \neq 1} \text{rand}_i F_{ij}^d(t), \tag{2.10}$$

where K_{best} is a number referred to the best agents and rand is a random number in the range from 0 to 1.

2.5 Related Work

Since PSO was introduced, many improvement and applications have been made and some of them related to our work are described below.

Chunshien and Tsunghan in 2011 [12] use PSO to construct fuzzy set and recursive least square estimator to calculate the fuzzy rules, in an approach for tuning a fuzzy system used in function approximation. This work is an application of PSO to optimize a fuzzy system.

Muthukaruppan and Er in 2012 [13] apply PSO along with a fuzzy expert system for diagnosis of coronary artery disease, in an effort to demonstrate the efficiency of PSO to find great solutions in the search space. This work is a hybridization of PSO with a fuzzy system.

Hongbo and Ajith in 2007 [14] proposed a new parameter in the velocity equation of PSO as a velocity threshold used to control the velocity of the particles, with a fuzzy logic controller used to adaptively control the new parameter. This work is related to our work because of the use of a fuzzy system to control a parameter in PSO, even when it is a new parameter.

Shi and Eberhart in 2001 [15] use a fuzzy system to adjust the inertia weight in PSO, which uses the current best performance evaluation and the current inertia weight to adjust the inertia weight. This work is related to our work because of the use of a fuzzy system to adjust a parameter in PSO, but using other inputs for the fuzzy system.

Taher et al. in 2012 [16] use a fuzzy system to adjust c_1, c_2 and w of PSO applied to distribution feeder reconfiguration. This work is related to our work because of the use of a fuzzy system to adjust three parameters in PSO, but using a different variant of PSO.

Wang et al. in 2006 [17] use a fuzzy system to change the velocity of the particles using the distance between the particles and decide if it is necessary to change their velocity in a violent way, to improve the global search of PSO. This work is related to our work because they use a fuzzy system to enhance the abilities of PSO to keep searching, but involves a direct operation with the particles and not through the parameters of PSO.

ACO has had many improvements and some of them are described below.

Neyoy et al. in 2012 [22] proposed an improvement in ACO with dynamic parameter tuning, using a fuzzy controller for α (alpha) parameter, using an average λ (lambda) branching factor reference level. This work is a previous work in ACO where they use a fuzzy system to control just one parameter.

Van Ast et al. in 2009 [21] use ACO with a fuzzy partitioning of the state space to the automated design of control policies in a dynamic system. This work presents a hybridization of ACO and a fuzzy system.

Yu et al. in 2012 [18] proposed an improved pheromone laying mechanism, where they change the way in which ants deposit the pheromone, changing the amount of ants that deposit the pheromone over the iterations. This work presents an improvement in ACO but without the use of a fuzzy system.

Einipour in 2011 [19] proposed a fuzzy ACO method to detect breast cancer, where ACO as evolutionary algorithm is applied to optimize a set of fuzzy rules used to classify cancer instances; these fuzzy rules are extracted from a breast cancer diagnosis dataset. This work is an application of ACO to optimize a fuzzy system.

Elloumi et al. in 2013 [20] proposed fuzzy PSO and fuzzy ACO as a hybridization applied to TSP, where they use fuzzy systems to update the inertia weight in PSO and in ACO for the weighting coefficient of the pheromone trails update. This work is related to our work because of the use of a fuzzy system to update a parameter in PSO and in ACO.

Khan and Engelbrecht in 2008 [23] use a fuzzy system to handle the multi-objective aspect of the topology design in distributed local area networks optimized with ACO. This work is an application of ACO along with a fuzzy system.

As related works with GSA, some of them are described below.

Sombra et al. in 2013 [24] use a fuzzy system for parameter adaptation of α (alpha) parameter in the gravitational search algorithm. This work is a previous work in GSA using a fuzzy system to adapt just one parameter.

Hassanzadeh and Rouhani in 2010 [25] proposed a multi-objective gravitational search algorithm, where GSA is changed to be applied in multi-objective problems and present a comparison with a multi-objective particle swarm optimization. This work presents an improvement in GSA to be multi-objective.

Mirjalili and Hashim in 2010 [26] proposed a hybridization of PSO and GSA for function optimization, as an integration of the exploitation of PSO and the exploration of GSA. This work presents a hybridization of PSO and GSA.

There are other works related to this book like: an optimization of fuzzy controllers in [27], an application of GA and PSO in [28, 29], an application of PSO in [30], and the design and optimization of fuzzy controllers in [31].

References

1. Zadeh L (1965) Fuzzy sets. Inf Control 8(3):338–353
2. Zadeh L (1965) Fuzzy logic. IEEE Comput 21(4):83–93
3. Zadeh L (1975) The concept of a linguistic variable and its application to approximate reasoning—I. Inf Sci 8:199–249
4. Liang Q, Mendel J (2000) Interval type-2 fuzzy logic systems: theory and design. IEEE Trans Fuzzy Syst 8(5):535–550
5. Mendel J, John R (2002) Type-2 fuzzy sets made simple. IEEE Trans Fuzzy Syst 10(2):117–127
6. Kennedy J, Eberhart R (1995) Particle swarm optimization. In: Proceedings of IEEE international conference on Neural Networks, IV. IEEE Service Center, Piscataway, NJ, pp 1942–1948
7. Kennedy J, Eberhart R (2001) Swarm intelligence. Morgan Kaufmann, San Francisco
8. Engelbrecht A (2006) Fundamentals of computational swarm intelligence. Wiley, Hoboken
9. Haupt R, Haupt S (1998) Practical genetic algorithms, 2nd edn. Wiley-Interscience, New York

10. Dorigo M (1992) Optimization, learning and natural algorithms. PhD Thesis, Dipartimento di Elettronica, Politechico di Milano, Italy
11. Rashedi E, Nezamabadi-pour H, Saryazdi S (2009) GSA: a gravitational search algorithm. Inf Sci 179(13):2232–2248
12. Chunshien L, Tsunghan W (2011) Adaptive fuzzy approach to function approximation with PSO and RLSE. Exp Syst Appl 38:13266–13273
13. Muthukaruppan S, Er M (2012) A hybrid particle swarm optimization based fuzzy expert system for the diagnosis of coronary artery disease. Exp Syst Appl 39:11657–11665
14. Hongbo L, Ajith A (2007) A fuzzy adaptive turbulent particle swarm optimization. Int J Innov Comput Appl 1(1):39–47
15. Shi Y, Eberhart R (2001) Fuzzy adaptive particle swarm optimization. In: Proceeding of IEEE international conference on evolutionary computation, IEEE Service Center, Piscataway, NJ, Seoul, Korea, pp 101–106
16. Taher N, Ehsan A, Masoud J (2012) A new hybrid evolutionary algorithm based on new fuzzy adaptive PSO and NM algorithms for distribution feeder reconfiguration. Energy Convers Manag 54:7–16
17. Wang B, Liang G, ChanLin W, Yunlong D (2006) A new kind of fuzzy particle swarm optimization fuzzy_PSO algorithm. In: 1st international symposium on systems and control in aerospace and astronautics, ISSCAA, pp 309–311
18. Neyoy H, Castillo O, Soria J (2012) Dynamic fuzzy logic parameter tuning for ACO and its application in TSP problems. SCI 451:259–271
19. Van Ast J, Babuska R, De Schutter B (2009) Fuzzy ant colony optimization for optimal control. In: Proceedings of the 2009 American control conference, St. Louis, Missouri, pp 1003–1008
20. Yu L, Yan JF, Yan GR, Yi L (2012) ACO with fuzzy pheromone laying mechanism. In: Emerging intelligent computing technology and applications. Springer, Berlin
21. Einipour A (2011) A fuzzy-ACO method for detect breast cancer. Glob J Health Sci 3(2):195
22. Elloumi W, Baklouti N, Abraham A, Alimi AM Hybridization of fuzzy PSO and fuzzy ACO applied to TSP. In: Hybrid intelligent systems (HIS), 2013 13th International Conference. IEEE, pp 105–110
23. Khan SA, Engelbrecht AP (2008) A fuzzy ant colony optimization algorithm for topology design of distributed local area networks. In: Swarm intelligence symposium. SIS 2008. IEEE, pp 1–7
24. Sombra A, Valdez F, Melin P, Castillo O (2013). A new gravitational search algorithm using fuzzy logic to parameter adaptation. In: 2013 IEEE congress on evolutionary computation (CEC), pp 1068–1074
25. Hassanzadeh HR, Rouhani M (2010) A Multi-objective gravitational search algorithm. In IEEE: second international conference on computational intelligence, communication systems and networks (CICSyN), Liverpool, pp 7–12
26. Mirjalili S, Hashim SZM (2010) A new hybrid PSOGSA algorithm for function optimization. In: IEEE: international conference on computer and information application (ICCIA), Tianjin, pp 374–377
27. Chandra SP, Amin MF, Akhand MAH, Murase K (2012) Optimization of interval type-2 fuzzy logic controller using quantum genetic algorithms. In: IEEE world congress on computational intelligence, pp 1027–1034
28. Oha S-K, Janga H-J, Pedrycz W (2011) A comparative experimental study of type-1/type-2 fuzzy cascade controller based on genetic algorithms and particle swarm optimization. Exp Syst Appl 38(9):11217–11229
29. Martinez R, Rodriguez A, Castillo O, Aguilar LT (2010) UABC, Tijuana, Mexico. Type-2 fuzzy logic controllers optimization using genetic algorithms and particle swarm optimization. In: 2010 IEEE international conference on granular computing (GrC). ISBN: 978-1-4244-7964-1
30. Al-Jaafreh MO, Al-Jumaily AA (2007) Training type-2 fuzzy system by particle swarm optimization. In: IEEE congress on evolutionary computation 2007, CEC 2007. ISBN: 978-1-4244-1339-3
31. Castillo O, Melin P (2012) A review on the design and optimization of interval type-2 fuzzy controllers. Appl Soft Comput 12(4):1267–1278

Chapter 3
Problem Statements

This chapter contains a description of the problems used to test the proposed methodology, like the benchmark mathematical functions, the travel salesman problem, the optimization of a trajectory of an autonomous mobile robot, and the fuzzy control of the temperature in a shower.

3.1 Benchmark Mathematical Functions

This type of problem consists in the minimization or to find the coordinates of the global minimum from a mathematical function with a complex surface, with a known global minimum.

A set of benchmark mathematical functions obtained from [1–3] are used to test the proposed methodology; in Fig. 3.1, a sample of the functions used is shown. Different dimensions are used in this type of problem but most of them are 30-dimensional benchmark mathematical functions.

In these problems, the objective in case of minimization is to find the lowest value of Z given X and Y (in a three-dimensional graph). If we try to find X and Y, this becomes in a two-dimensional problem, even when it can be represented in a three-dimensional space like the functions shown in Fig. 3.1.

3.2 Travel Salesman Problem

This problem consists in a salesman that needs to visit only once each one of the given cities; the salesman can start in any city and at the end return to the city from where he started. There are arcs that connect the cities and each one has a cost or distance. So the ideal tour is which it has the minimum cost or total distance. In other words, a solution to this problem is a Hamiltonian cycle with minimal cost.

© The Author(s) 2018   11
F. Olivas et al., *Dynamic Parameter Adaptation for Meta-Heuristic Optimization Algorithms Through Type-2 Fuzzy Logic*, SpringerBriefs in Computational Intelligence,
https://doi.org/10.1007/978-3-319-70851-5_3

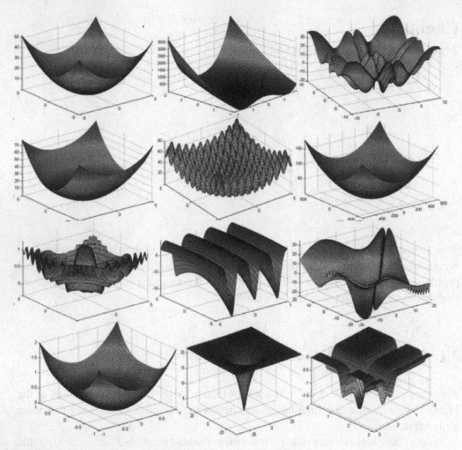

Fig. 3.1 Sample of the benchmark mathematical function used

Table 3.1 Characteristics of the TSP problems used

TSP	Number of cities	Minimum
Burma14	14	3323
Ulysses22	22	7013
Berlin52	52	7542
Eil76	76	538
KroA100	100	21,282

In Fig. 3.2, a graphical representation of a TSP with 100 cities is illustrated, with its global minimum tour.

Five problems were chosen from the TSPLIB library [4], with different number of cities involved in each problem. Table 3.1 contains the characteristics of each of these problems.

The first problem is the easiest one, called Burma14, and has only 14 cities, and the minimum tour has a length of 3323. Ulysses22 has 22 cities from the Odyssey

Fig. 3.2 Graphical representation of a TSP

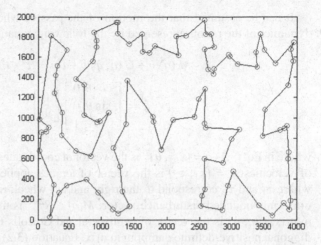

of Ulysses, and the minimum tour has a length of 7013. Berlin52 has 52 places in Berlin, and the minimum tour has a length of 7542. Eil76 has 76 cities, and the minimum tour has a length of 538. KroA100 has 100 cities, and the minimum tour has a length of 21,282.

3.3 Autonomous Mobile Robot

The robot used is a wheeled vehicle capable of moving in different environments; its body is symmetrical and has two wheeled motors and one passive wheel that provides stabilization, to prevent that the robot falls over while it moves on an environment (Fig. 3.3).

Fig. 3.3 Scheme of the wheeled mobile robot

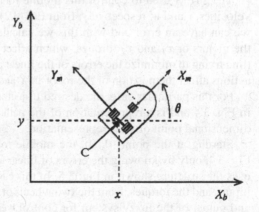

It can be assumed that the motion of the passive wheel can be ignored in the dynamics of the robot represented by the following equations [5]:

$$M(q)\dot{v} + C(q,\dot{q})v + Dv = \tau + P(t) \tag{3.1}$$

$$\dot{q} = \begin{bmatrix} \cos\theta & 0 \\ \sin\theta & 0 \\ 0 & 1 \end{bmatrix} \begin{bmatrix} v \\ w \end{bmatrix}, \tag{3.2}$$

where in Eq. (3.1) $q = (x, y, \theta)^T$ is the vector of coordinates; $v = (v, w)^T$ is the vector of velocities; $\tau = (\tau_1, \tau_2)$ is the vector of torques applied to the wheeled motors, where τ_1 and τ_2 correspond to the right and left wheel, respectively; $P \in R^2$ is the uniformly bounded disturbance vector; $M(q) \in R^{2\times2}$ is the positive-definite inertia matrix; $C(q, \dot{q})$ is the vector of centripetal and Coriolis forces; and $D \in R^{2\times2}$ is a diagonal positive-definite damping matrix. Equation (3.2) represents the kinematics of the system, θ is the angle between the heading direction and the x-axis; v and w are the linear and angular velocities, respectively. Furthermore, the system of Eqs. (3.1) and (3.2) has the following non-holonomic constraint:

$$\dot{y}\cos\theta - \dot{x}\sin\theta = 0 \tag{3.3}$$

Equation (3.3) corresponds to a condition to prevent that the robot moves sideways from [6]. The system of Eq. (3.2) fails to meet Brockett's necessary condition for feedback stabilization from [7], which implies that no continuous static state-feedback controller exists that stabilizes the closed-loop system around the equilibrium point.

The objective is to design a fuzzy controller of τ that ensures

$$\lim_{t \to \infty} \|q_d(t) - q(t)\| = 0 \tag{3.4}$$

for any desired trajectory $q_d \in R^3$ while attenuating external disturbances.

The T1FIS used to control this mobile robot is based on the linear and angular velocities, v and w, respectively, from Eq. (3.2), and based on the desired velocities, we can have an error and with this we calculate the force that must be applied to the motors or τ_1 and τ_2 torques, which affect directly the wheels, and at the same time trying to minimize the errors of the linear and angular velocities, and with these actions affect the position of the robot in X and Y.

For this particular case, the desired trajectory of the mobile robot is represented in Fig. 3.4 and is a representation of the path that the robot must follow in a two-dimensional point of view representation.

Starting at the point (0, 0), the mobile robot must follow the reference from Fig. 3.4, only by knowing the errors of linear and angular velocities, and the T1FIS uses the structure shown in Fig. 3.5. In this case, the errors of the velocities are the inputs, and the torques, from the two wheels of the robot, are the outputs. Each input and output of the fuzzy system for control were granulated into three membership functions (Negative, Zero, and Positive).

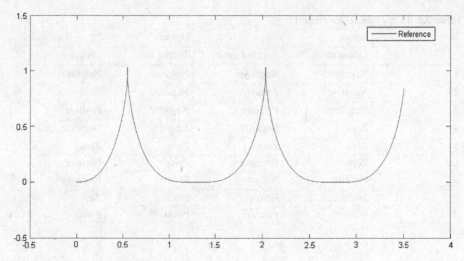

Fig. 3.4 Path of reference for the mobile robot

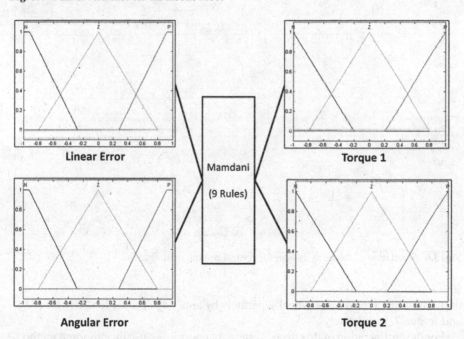

Fig. 3.5 Structure of the fuzzy system for control

The fuzzy rules used in the fuzzy controller of the mobile robot are listed in Table 3.2, and these rules were created based on a simple control strategy taking into account the linear and angular errors and their effect on the position of the robot, so

Table 3.2 Fuzzy rule set for control

No.	Inputs		Outputs	
	Linear error	Angular error	Torque 1	Torque 2
1	Negative	Negative	Negative	Negative
2	Negative	Zero	Negative	Zero
3	Negative	Positive	Negative	Positive
4	Zero	Negative	Zero	Negative
5	Zero	Zero	Zero	Zero
6	Zero	Positive	Zero	Positive
7	Positive	Negative	Positive	Negative
8	Positive	Zero	Positive	Zero
9	Positive	Positive	Positive	Positive

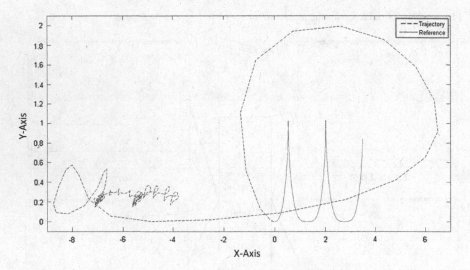

Fig. 3.6 Simulation of the fuzzy system for control without optimization

with this we can move each wheel separately by applying specific forces in torque 1 and torque 2, respectively.

For the optimization of this fuzzy system for control, we use the proposed method to search for the best points of each membership function, and the rule set stays with no changes. In addition, for comparison purposes, we apply this fuzzy system with no optimization (known as the basis fuzzy controller), to the control of the mobile robot and the simulation is presented in Fig. 3.6.

In Fig. 3.6, we can observe that the reference (green line) and the trajectory of the robot (blue line) start at the same point (0, 0) but the fuzzy controller gets lost immediately; and in the following section, we can find that with the proposed approach, and only with the optimization of the parameters of each membership function (the rule set has no change), the trajectory of the robot will be improved significantly.

3.4 Automatic Temperature Control in a Shower

The second problem is of a benchmark in control, where the control provides an automatic adjustment of the water temperature; the objective is to control the level of temperature and flow of the water in a shower using a fuzzy controller. All the methods are applied to optimize the points of the membership functions of the fuzzy system used to control the levels of hot and cold water.

Figure 3.7 illustrates the plant used to control the temperature in a shower; this is done with the help of a fuzzy system with which the levels of hot and cold water are controlled.

The fuzzy system used to control the levels of hot and cold water is illustrated in Fig. 3.8, where it has two inputs (temperature and flow) and two outputs (cold and hot), using a fuzzy set of nine rules shown in Table 3.3.

Table 3.3 contains the fuzzy rules for the fuzzy controller from Fig. 3.8, to control the levels of hot and cold water in a shower.

The fuzzy rule set from Table 3.3 represents the knowledge of the problem of controlling the levels of hot and cold water in a shower. The fuzzy system provides

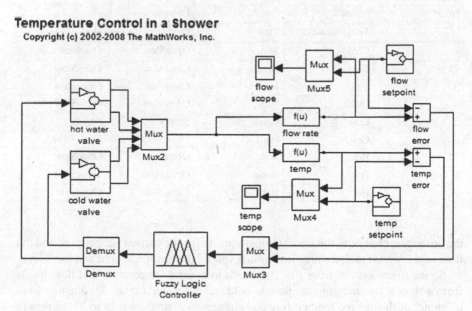

Temperature Control in a Shower
Copyright (c) 2002-2008 The MathWorks, Inc.

Fig. 3.7 Plant for temperature control in a shower

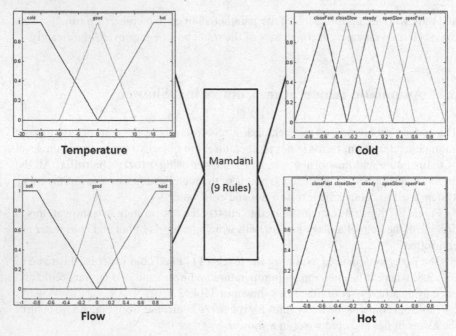

Fig. 3.8 Fuzzy controller to automatically adjust the temperature in a shower

Table 3.3 Fuzzy rules for the fuzzy controller

No.	Inputs		Outputs	
	Temperature	Flow	Cold	Hot
1	Cold	Soft	OpenSlow	OpenFast
2	Cold	Good	CloseSlow	OpenSlow
3	Cold	Hard	CloseFast	CloseSlow
4	Good	Soft	OpenSlow	OpenSlow
5	Good	Good	Steady	Steady
6	Good	Hard	CloseSlow	CloseSlow
7	Hot	Soft	OpenFast	OpenSlow
8	Hot	Good	OpenSlow	CloseSlow
9	Hot	Hard	CloseSlow	CloseFast

the automatic change in the water temperature, using the current temperature and the flow of the water, to determine how much will be the change in cold and hot water.

So the fuzzy system from Fig. 3.8 needs to control temperature and flow levels from a shower by changing the flow in cold and hot water. Figure 3.9 illustrates the reference of the desired temperature and changes over time from 19 to 27 degrees in temperature.

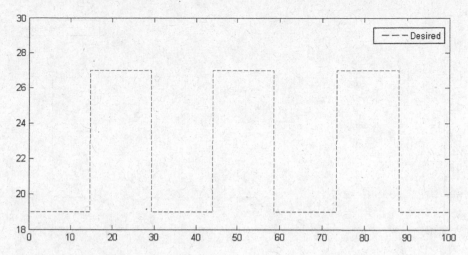

Fig. 3.9 Reference of the desired temperature

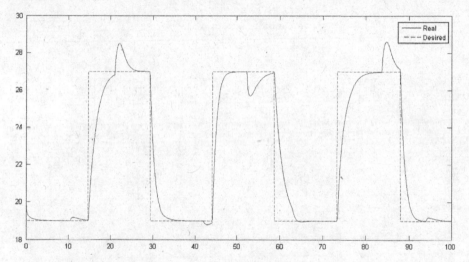

Fig. 3.10 Result in the temperature reference of applying the fuzzy controller used as base

With the fuzzy controller illustrated in Fig. 3.8 and using the fuzzy rule set from Table 3.3 following the reference of the desired temperature from Fig. 3.9, the result is illustrated in Fig. 3.10.

Figure 3.11 illustrates the desired flow in the shower, its variable likes the temperature reference and changes from 0.5 to 0.9 every 10 units of time.

The result of the fuzzy controller in trying to follow the reference flow is illustrated in Fig. 3.12.

With both results from Figs. 3.10 and 3.12, the results of applying the fuzzy controller to the automatic temperature control in a shower is an MSE = 2.7758.

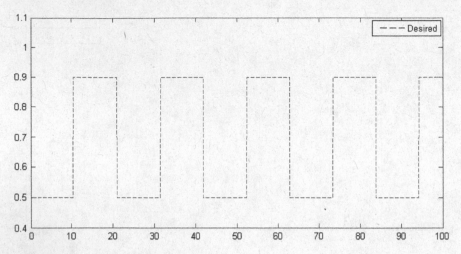

Fig. 3.11 Reference of the desired flow

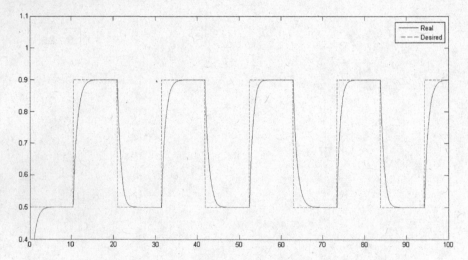

Fig. 3.12 Result in the reference flow of applying the fuzzy controller used as a basis

References

1. Haupt R. L, Haupt S. E (1998) Practical genetic algorithms (Vol. 2). New York, Wiley
2. Marcin M, Smutnicki C (2005) Test functions for optimization needs. http://www.
   bioinformaticslaboratory.nl/twikidata/pub/Education/NBICResearchSchool/Optimization/
   VanKampen/BackgroundInformation/TestFunctions-Optimization.pdf
3. Rashedi E, Nezamabadi-pour H, Saryazdi S (2009) GSA: a gravitational search algorithm. Inf
   Sci 179(13):2232–2248

4. Reinelt G (1991) TSP_LIB—A traveling salesman problem library. ORSA J Comput 3:376–384
5. Lee T-C, Song K-T, Lee C-H, Teng C-C (2001) Tracking control of unicycle-modeled mobile robot using a saturation feedback controller. IEEE Trans Control Syst Technol 9:305–318
6. Liberzon D (2003) Switching in systems and control. Birkhauser, Basel
7. Brockett RW (1983) Asymptotic stability and feedback stabilization. In: Millman RS, Sussman HJ (eds) Differential geometric control theory. Birkhauser, Boston, pp 181–191

Chapter 4
Methodology

In this chapter, the proposed methodology is described in detail, in other words, how the parameters to be adjusted are selected, how to create the fuzzy system for parameter adaptation, and the creation of the fuzzy rules based on knowledge of the behavior of the optimization algorithm.

4.1 Description of the Methodology

The proposed methodology was designed to be easily included in an optimization algorithm but the algorithm needs to meet some restrictions like the optimization algorithm needs to be iterative, and the Euclidean distance can be calculated from the individuals of the population.

The dynamic parameter adaptation of the optimization algorithm is through a fuzzy system used to control the behavior and model a better performance in the execution of itself. Our methodology uses the iterations from the optimization algorithm to model the behavior, that is, in early iterations the algorithm needs to explore the entire search space, but in final iterations the algorithm needs to exploit the best area found during the iterations; basically, that is the reason for the need of the iterations in the optimization algorithm.

To help the iterations, a metric about the distribution of the population in the search space is needed. So the diversity of the individuals is needed, and to calculate the diversity it is necessarily the Euclidean distance.

Once the metrics are defined, it is necessary to perform a brief study of the parameters of the algorithm, and with this study select the most important parameters, the parameters which have the most impact in the behavior of the algorithm, in order to control the way in which the algorithm performs the global or the local search.

© The Author(s) 2018
F. Olivas et al., *Dynamic Parameter Adaptation for Meta-Heuristic Optimization Algorithms Through Type-2 Fuzzy Logic*, SpringerBriefs in Computational Intelligence, https://doi.org/10.1007/978-3-319-70851-5_4

Also, from the brief study of the parameters of the algorithm, we can know the behavior of the parameters in the algorithm and extract this knowledge into the fuzzy rule set, used by the fuzzy system for parameter adaptation.

4.2 General Procedure of the Methodology

The proposed methodology is designed with two basic inputs, the iteration and diversity described in Sect. 4.3, and these metrics work fine, as mentioned in Sect. 1.4 and in the experiment section of this book but other metrics can be used, i.e., how many iterations are elapsed since the last change in the best value found yet, a metric designed specifically for an optimization method, or the difference in fitness between the best individual and the population, etc.

To use the proposed methodology in an optimization algorithm, a general procedure is described below, using the two metrics of iteration and diversity described in Sect. 4.3.

1. The optimization method needs to be iterative, in order to use the iteration metric.
2. The Euclidean distance can be calculated between the individuals, in order to use the diversity metric.
3. A study of parameters user-defined from the optimization method needs to be done, to know the behavior of the algorithm and the impact that has every parameter lonely. This study is experimental to define the best range of values for a parameter and to see the effect of changing a parameter.
4. The parameters with the most impact in the behavior of the algorithm are the most important parameters.
5. From the study of parameters, a fuzzy rule set can be obtained to control the parameters.
6. Select the best place in the optimization algorithm to include the fuzzy system for parameter adaptation, usually the best place is just before the use of the parameter, and between the main loops of iterations.
7. For each parameter selected to be adjusted, create a type-1 (or an interval type-2) fuzzy system with only one input and one output to see if the adaptation of that parameter brings better quality results.
8. Combine all the fuzzy systems with one input and one output to create a more complex fuzzy system and obtain even better quality results.

With the inclusion of the proposed methodology, an optimization method can obtain better quality results but the use of a fuzzy system needs more computational time, so the proposed methodology is ideal to use when the time is not a critical issue, because it can offer a better result.

4.3 Metrics Used in the Methodology

The iteration metric is very helpful in the design of a better behavior of the optimization algorithm, because we need that in early iterations perform a global search, and in final iterations perform a local search, but as reinforcement we use the diversity metric to know how close the individuals from each other are.

The iteration metric is calculated using Eq. (4.1) and is basically a percentage of the elapsed iterations.

$$\text{Iteration} = \frac{\text{Current iteration}}{\text{Maximum of iterations}}, \tag{4.1}$$

where the *current iteration* is the actual iteration, and the *maximum of iterations* is the total number of iterations established for the optimization algorithm to perform its search.

The diversity metric is the degree of dispersion of all of the individuals from the best and is calculated using Eq. (4.2), and can be seen as an average of the Euclidean distance between the entire population and the best.

$$\text{Diversity}(S(t)) = \frac{1}{n_s} \sum_{i=1}^{n_s} \sqrt{\sum_{j=1}^{n_x} \left(X_{ij}(t) - \bar{X}_j(t)\right)^2}, \tag{4.2}$$

where *diversity* is a degree of dispersion of the population, S is the population, n_s is the number of individuals in the population, n_x is the number of dimensions of the problem, x_{ij} is the position of an individual, and *x-bar* is the position of the best individual.

The metrics are the inputs for the fuzzy system used for parameter adaptation; Fig. 4.1 illustrated the iteration metric, where the range of the input variable iteration is from 0 to 1 and is granulated into three triangular membership functions (Low, Medium, and High), and in Fig. 4.2 the diversity metric is illustrated, where the input variable diversity has a range from 0 to 1 and granulated into three triangular membership functions (Low, Medium, and High).

The fuzzy rule set depends on the knowledge of the behavior that we want to model from each optimization method and the parameters selected to be adjusted dynamically over the iterations.

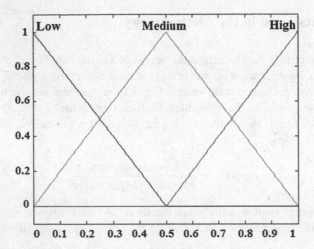

Fig. 4.1 Iteration metric as input for the fuzzy system

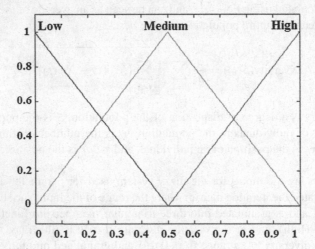

Fig. 4.2 Diversity metric as input for the fuzzy system

4.4   Selection of Parameters to Be Adjusted

Each optimization method has its own parameters and depends on the problem in which it is applied, that is, in a type of problem a parameter needs to be low but for another problem the same parameter would need to be high. This adjustment of parameters is commonly made manually and requires several experiments to find the best values for each parameter; the problem is that this search for the best values needs to be done for each type of problem. Also, each optimization method has some parameters that have the most impact on its behavior.

Table 4.1 Fuzzy rules for the parameter adaptation in PSO

No.	Inputs		Outputs		
	Iteration	Diversity	c_1	c_2	Constriction
1	Low	Low	High	Low	High
2	Low	Medium	MediumHigh	MediumLow	MediumHigh
3	Low	High	MediumHigh	MediumLow	Medium
4	Medium	Low	MediumHigh	MediumLow	MediumHigh
5	Medium	Medium	Medium	Medium	Medium
6	Medium	High	MediumLow	MediumHigh	MediumLow
7	High	Low	Medium	High	Medium
8	High	Medium	MediumLow	MediumHigh	MediumLow
9	High	High	Low	High	Low

Our methodology performs the adjustment of parameters dynamically and only depends on the metrics described above but only requires once the experimentation to know the behavior of the optimization method and with this create the fuzzy rule set to control the parameters. Basically, the desired behavior of any optimization method is: at the beginning is necessary a global search, and at the end is necessary a local search.

For PSO using knowledge about the effects of the parameters in the velocity Eq. (2.2), for example, when PSO uses a c_1 higher than c_2, the effect is that the swarm will explore the search space, and when it uses a c_1 lower than c_2, the effect is that the swarm will exploit the best area of the search space found so far. The effect of the constriction factor is to multiply the new velocity of the particle, so with a constriction factor large the swarm will explore the search space, and with a constriction factor low the swarm will exploit the best area of the search space found so far. Table 4.1 described the fuzzy rule set used in PSO to adapt dynamically the c_1 and c_2 parameters and the constriction factor.

With the fuzzy rule set in Table 4.1, an improved behavior of PSO is designed, so in early iterations PSO uses exploration or performs a global search and in final iterations uses exploitation or performs a local search.

For ACO based on experiments, the parameters α (alpha) from Eq. (2.3) and ρ (rho) from Eq. (2.4) were selected to be adjusted dynamically; in early iterations, α needs to be low and ρ needs to be high to increase the global search, and in final iterations, α needs to be high and ρ needs to be low to perform a local search. The knowledge of the behavior of ACO is used to create the fuzzy rule set from Table 4.2.

The fuzzy rule set from Table 4.2 is designed to provide a desired behavior to ACO, in order to control the abilities of ACO and perform a global search in early iterations and perform a local search in final iterations.

Table 4.2 Fuzzy rules for the parameter adaptation in ACO

No.	Inputs		Outputs	
	Iteration	Diversity	Alpha α	Rho ρ
1	Low	Low	Low	High
2	Low	Medium	MediumLow	MediumHigh
3	Low	High	Medium	Medium
4	Medium	Low	MediumLow	MediumHigh
5	Medium	Medium	Medium	Medium
6	Medium	High	MediumHigh	MediumLow
7	High	Low	Medium	Medium
8	High	Medium	MediumHigh	MediumLow
9	High	High	High	Low

Table 4.3 Fuzzy rules for the parameter adaptation in GSA

No.	Inputs		Outputs	
	Iteration	Diversity	Alpha α	*Kbest*
1	Low	Low	Low	High
2	Low	Medium	MediumLow	MediumHigh
3	Low	High	Medium	Medium
4	Medium	Low	MediumLow	MediumHigh
5	Medium	Medium	Medium	Medium
6	Medium	High	MediumHigh	MediumLow
7	High	Low	Medium	Medium
8	High	Medium	MediumHigh	MediumLow
9	High	High	High	Low

For GSA, the parameters α (alpha) from Eq. (2.9) and *Kbest* from Eq. (2.10) were selected to be adjusted dynamically over the iterations of the optimization algorithm. From the knowledge of the behavior of the parameters in GSA, the fuzzy rule set from Table 4.3 was created.

The parameters selected for each method were the ones which have the most impact in the behavior from each optimization method, that is, each parameter was changed in each experiment from a recommended range of values and studied the behavior of the optimization method.

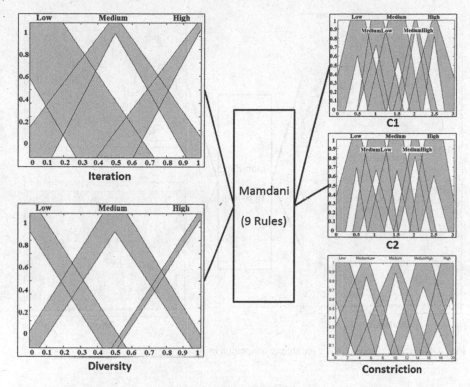

Fig. 4.3 Final fuzzy system for parameter adaptation in PSO

4.5 Final Fuzzy Systems for Parameter Adaptation

The fuzzy systems used for parameter adaptation in the different methods were evolved from a simple fuzzy system with one input and one output, to an interval type-2 fuzzy system with two inputs (iteration and diversity) and two (or three) outputs.

For PSO, the final fuzzy system is illustrated in Fig. 4.3, where an interval type-2 fuzzy system is used to adapt c_1, c_2 and *constriction factor* from Eq. (2.2), using the fuzzy rule set from Table 4.1.

From Fig. 4.3, we can see that the iteration and diversity inputs were granulated into three-interval type-2 triangular membership functions with a range from 0 to 1, and the outputs were granulated into five-interval type-2 triangular membership functions, c_1 and c_2 with a range from 0 to 3 and the *constriction factor* from 0 to 20.

For ACO, the final fuzzy system is shown in Fig. 4.4, used for the adjustment of α (*alpha*) and ρ (*rho*) from Eqs. (2.3) to (2.4), respectively, and uses the fuzzy rule set from Table 4.2.

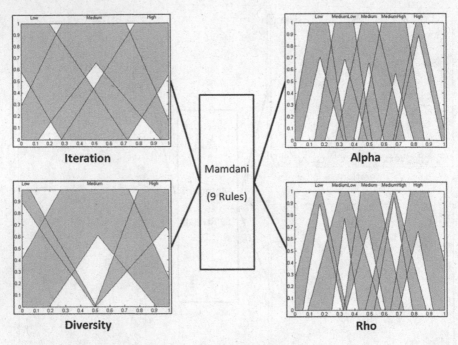

Fig. 4.4 Final fuzzy system for parameter adaptation in ACO

The inputs and outputs have a range from 0 to 1 illustrated in Fig. 4.4 but the inputs were granulated into three-interval type-2 triangular membership functions, and the outputs were granulated into five-interval type-2 triangular membership functions.

For the GSA algorithm, the final fuzzy system for the adaptation of the α (*alpha*) and *Kbest* parameters from Eqs. (2.9) and (2.10), respectively, is illustrated in Fig. 4.5 and we use the fuzzy rule set from Table 4.3.

The inputs *iteration* and *diversity* and the output *Kbest* have a range from 0 to 1 illustrated in Fig. 4.5, and the output α (*alpha*) has a range from 0 to 100; the inputs are granulated into three-interval type-2 triangular membership functions, and the outputs are granulated into five-interval type-2 triangular membership functions.

As mentioned early, when a new optimization method is selected to create a fuzzy system for parameter adaptation, first we only develop type-1 fuzzy systems with one input and one output, and this is because it is easier to find the fuzzy rules when it only has one input and one output, and eventually try to combine those fuzzy systems into one more complex with two inputs and two or three outputs.

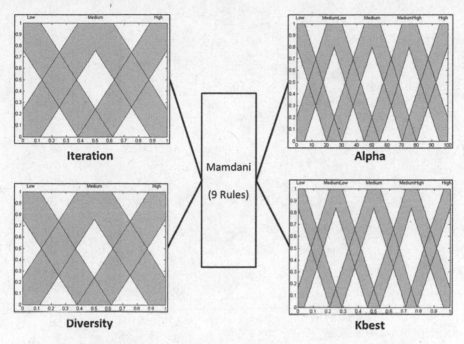

Fig. 4.5   Final fuzzy system for parameter adaptation in GSA

In all of the optimization algorithms used to control their parameters, the use of the input iteration helps in obtaining better quality results, but once the input diversity is added, it has even better results, and this is because with these two inputs, we can have a better control of the parameters in an optimization algorithm; therefore, controlling their parameter, we can control their behavior and its abilities to perform a global or a local search.

Chapter 5
Simulation Results

This chapter shows a detailed description of the results of the problems described in Chap. 3, with all of the optimization methods modified with the proposed methodology and compared with the original methods to see the difference in results.

5.1 Results with the Benchmark Mathematical Functions

The first problem to test the proposed methodology is the one of the benchmark mathematical functions described in Sect. 3.1, where we use two set of functions one of 26 from [1, 2], and one of 15 from [3], most of these functions are used with 30 dimensions with the same constrictions for all the methods.

Also, a comparison with other papers is presented with the same benchmark mathematical functions and using the same constrictions used on each paper.

Table 5.1 contains the results with the 26 benchmark mathematical functions using the original PSO method and the best version of PSO developed with the proposed methodology, with the parameters *population* = 30, *dimensions* = 30, and *iterations* = 100. In this case from Table 5.1, original PSO has a decreasing inertia weight and FPSO has a dynamic parameter adaptation of c_1, c_2 and constriction factor from Eq. (2.2), using the interval type-2 fuzzy system from Fig. 4.3 and the fuzzy rule set from Table 4.1. It is important to emphasize that each result from Table 5.1 is an average from 30 experiments; this method was applied to the same function 30 times and an average of the results is presented; also, the results in bold are the best from each function.

From Table 5.1, the results show the improvement in PSO, where FPSO with the dynamic parameter adaptation can achieve better quality results when compared with the original PSO method. On average, the proposed improvement in PSO obtains better results in all the benchmark mathematical functions.

The same FPSO method from Table 5.1 was used to perform some comparisons with specific functions, described in the papers used in the comparison.

© The Author(s) 2018

F. Olivas et al., *Dynamic Parameter Adaptation for Meta-Heuristic Optimization Algorithms Through Type-2 Fuzzy Logic*, SpringerBriefs in Computational Intelligence, https://doi.org/10.1007/978-3-319-70851-5_5

Table 5.1 Results with the 26 benchmark mathematical functions

Function	Original PSO	FPSO with parameter adaptation
F1	4.6680e−4	1.2568e−9
F2	8.0410e−7	3.1924e−18
F3	1.3661	0.0046
F4	1.4917e−5	4.3597e−13
F5	0.6890	0.0365
F6	0.1150	0.0014
F7	0.1756	0.0095
F8	1.4686e−4	1.4802e−10
F9	0.0193	2.3359e−5
F10	3.3469e−5	2.0149e−12
F11	3.4586e−4	1.0019e−6
F12	0.0109	4.8791e−7
F13	0.0035	1.7682e−4
F14	0.2331	5.7417e−4
F15	0.0645	8.5165e−5
F16	0.1429	2.8497e−16
F17	3.3866e−6	5.1684e−12
F18	0.4535	4.3186e−8
F19	0.0255	4.8457e−5
F20	0.1691	2.3213e−4
F21	0.0220	3.5686e−5
F22	0.1363	5.1684e−15
F23	3.5199e−6	8.4849e−10
F24	4.0243e−4	4.8469e−6
F25	0.0159	4.6543e−4
F26	0.0550	8.4954e−5

The first comparison is against Hongbo and Ajith [4], where they add a new parameter to the velocity equation of PSO called minimum velocity threshold, which is used to control the velocity of particles and a fuzzy controller is used to tune adaptively this new parameter. To make the comparison, we use the same constraints as they used in [4], which are as follows: *population size* = 20, *iterations* = 18,000, and *dimensions* = 30 or 100. Table 5.2 summarizes the results presented in [4], and we add our results using the proposed approach, each result is an average of 10 experiments and results in bold are the best.

From Table 5.2, the proposed methodology for parameter adaptation in PSO can achieve better results and, when compared with another improvement in PSO, can outperform the results and on average is better.

The second comparison is against Shi and Eberhart [5], where they use a fuzzy system to adjust the inertia weight in PSO. Table 5.3 contains the results of the com-

Table 5.2 Comparison against Hongbo and Ajith [4]

Function	Dim	Hongbo et al. [4]	FPSO with parameter adaptation
Rosenbrock	30	1.1048e−4	2.0054e−5
	100	6.9026e−4	1.9862e−4
Schwefel	30	5.9520e−6	2.3568e−6
	100	9.2702e−4	8.4516e−4
Rastrigin	30	8.4007e−10	2.1354e−10
	100	19.9035	1.1954e−03
Griewank	30	0.0102	2.4492e−08
	100	0.0720	5.8496e−10
Ackley	30	5.4819e−4	2.6768e−19
	100	0.0011	1.3715e−04

Table 5.3 Comparison against Shi and Eberhart [5]

Rosenbrock

Population size	Dimensions	Iterations	Shi and Eberhart [5]	FPSO with parameter adaptation
20	10	1000	66.01409	2.5431
	20	1500	108.2865	15.8462
	30	2000	183.8037	38.8491
40	10	1000	48.76523	1.4956
	20	1500	63.88408	9.5468
	30	2000	175.0093	27.2133
80	10	1000	15.81645	2.5656
	20	1500	45.99998	8.5468
	30	2000	124.4184	22.8496

parison with the Rosenbrock function; Table 5.4 contains the results of the comparison with the Rastrigin function; and Table 5.5 contains the results of the comparison with the Griewank function. All of these results and tables are performed using the same constrictions as in [5], each result in Tables 5.3, 5.4 and 5.5 is an average of 50 experiments, and results in bold are the best.

From Table 5.3, the proposed methodology helps PSO to achieve better results in all the experiments, and on average, FPSO with parameter adaptation is better when compared with Shi and Eberhart [5], when is applied to the minimization of the Rosenbrock function.

From Table 5.4 in all the experiments, the proposed methodology helps PSO to achieve on average better results than Shi and Eberhart [5], when applied to the minimization of the Rastrigin function.

Table 5.4 Comparison against Shi and Eberhart [5]

Rastrigin

Population size	Dimensions	Iterations	Shi and Eberhart [5]	FPSO with parameter adaptation
20	10	1000	4.955165	1.8491
	20	1500	23.27334	5.0213
	30	2000	48.47555	16.8492
40	10	1000	3.283368	1.1334
	20	1500	15.04448	5.8704
	30	2000	35.20146	12.2354
80	10	1000	2.328207	0.5849
	20	1500	10.86099	3.2136
	30	2000	22.52393	9.6935

Table 5.5 Comparison against Shi and Eberhart [5]

Griewank

Population size	Dimensions	Iterations	Shi and Eberhart [5]	FPSO with parameter adaptation
20	10	1000	0.091623	0.0654
	20	1500	0.027275	0.0192
	30	2000	0.02156	0.0114
40	10	1000	0.075674	0.0568
	20	1500	0.031232	0.0149
	30	2000	0.012198	0.0097
80	10	1000	0.068323	0.0628
	20	1500	0.025956	0.0181
	30	2000	0.014945	0.0110

From Table 5.5, the results show that the proposed methodology for parameter adaptation in PSO helps in achieving better quality results and in all the experiments on average is better when is applied to the minimization of the Griewank function.

The third comparison is against Wang et al. [6], where a fuzzy system is used to determine when to change the velocity of some particles in PSO, using the distance between all the particles. Table 5.6 summarize the results from [6] and our results with the same constrictions, $population = 25$, $dimensions = 200$, $iterations = 20,000$, and an *asymmetric initialization*; the results in Table 5.6 are the best and results in bold are the best from each function.

From the results in Table 5.6, the proposed methodology for parameter adaptation in PSO can help in obtaining better quality results, with the same constraints as in [6].

Table 5.6 Comparison against Wang et al. [6]

Function	Wang et al. [6]	FPSO with parameter adaptation
Rosenbrock	8.1	0.1354
Rastrigin	8.6	0.1562

Table 5.7 Results with the set of 15 benchmark mathematical functions

Function	Original GSA from [3]	PSO from [3]	Sombra et al. [10]	FGSA with parameter adaptation
F1	7.3e−11	1.8e−3	8.8518e−34	5.4441e−42
F2	4.03e−5	2.0	1.1564e−10	6.3418e−24
F3	0.16e−3	4.1e+3	468.4431	2.2559e−7
F4	3.7e−6	8.1	0.0912	9.4936e−16
F5	25.16	3.6e+4	61.2473	1
F6	8.3e−11	1.0e−3	0.1000	0
F7	0.018	0.04	0.0262	1.00e−5
F8	−2.8e+3	−9.8e+3	−2.6792e+3	−3.6117e+3
F9	15.32	55.1	17.1796	2.5869e−3
F10	6.9e−6	9.0e−3	6.3357e−15	1.5704e−28
F11	0.29	0.01	4.9343	7.5050e−9
F12	0.01	0.29	0.1103	2.1135e−5
F13	3.2e−32	3.1e−18	0.0581	4.7641e−45
F14	3.70	0.998	2.8274	9.0713e−8
F15	8.0e−3	2.8e−3	0.0042	2.9205e−10

For PSO, our methodology shows an improvement and when compared with the original PSO method and even when compared with other improvements in PSO shows that our methodology is better and helps in obtaining better quality results in the minimization of benchmark mathematical functions.

The experimentation using GSA with parameter adaptation is shown in Table 5.7 with the set of 15 benchmark mathematical functions, where the original GSA method is applied to these functions and compared with the proposed modification of GSA with parameter adaptation, using the same constraints as in [3], with a *population* = 50, *iterations* = 1000 and *dimensions* = 30, also PSO from [3] has a decreasing inertia weight. The FGSA method from Table 5.7 has our methodology for parameter adaptation of α (*alpha*) from Eq. (2.9) and *Kbest* from Eq. (2.9), using the interval type-2 fuzzy system from Fig. 4.5 and using the fuzzy rule set from Table 4.3. Also, the results in bold in Table 5.7 are the best from each function, and each result is an average of 30 experiments.

The results in Table 5.7 show that the proposed methodology can also help GSA to obtain better quality results, when compared with the original GSA method and even when compared with an improvement in GSA through a fuzzy system and another optimization algorithm like PSO with inertia weight.

Table 5.8 Parameters of ACO for the experiments with TSP

Parameter	Original ACO	Neyoy et al. [7]	FACO with parameter adaptation
Iterations	1000	1000	1000
α (alpha)	1	Dynamic	Dynamic
β (beta)	2	2	2
ρ (rho)	0.1	0.1	Dynamic
Population	Number of cities	Number of cities	Number of cities

Table 5.9 Results in the experiments with TSP

Parameter	Minimum	Original ACO	Neyoy et al. [7]	FACO with parameter adaptation
Burma14	3323	3350.3	3323	3323
Ulysses22	7013	7089.1	7013	7013
Berlin52	7542	7850.4	7543	7542
Eil76	538	554.76	539	538
KroA100	21,282	22,591	21,344	21,369.63

In both methods, PSO and GSA, the proposed methodology for parameter adaptation helps in improvement of the performance and obtaining better quality results, when compared with the original methods, other improvements of the same optimization method and with different optimization methods.

5.2 Results with the Travel Salesman Problem

The second benchmark problem used to test the proposed methodology is the TSP described in Sect. 3.2, where the need to find the minimal tour in a graph is also described, and five different TSP (described in Table 3.1) are used in the experimentation with the original ACO method and ACO with adaptation.

In this case, the FACO method from Table 5.8 uses the proposed methodology for the adaptation of α (*alpha*) from Eq. (2.3) and ρ (*rho*) from Eq. (2.4), using the interval type-2 fuzzy system from Fig. 4.4, with the fuzzy rule set from Table 4.2, and the original ACO method is the rank-based variation. Neyoy et al. [7] proposed an improvement in ACO using a fuzzy system to control the α (*alpha*) parameter.

The results in Table 5.9 are using the parameters from Table 5.8, applied to the TSP from Table 3.1. Each result in Table 5.9 is an average of 30 experiments, and results in bold are the best.

From the results in Table 5.9, the first two problems are still a challenge for the original ACO method but easy for the methods with an improvement, in the case of

Table 5.10 Details from the results in Table 5.9

	Original ACO	Neyoy et al. [7]	FACO with parameter adaptation
Best	3323	3323	3323
	7013	7013	7013
	7543	7542	7542
	539	538	538
	22,798	21,292	21,282
Worst	3769	NP*	3323
	7258	NP*	7013
	8234	NP*	7542
	586	NP*	538
	23,609	NP*	22,114
Percentage in finding the minimum	12/30	30/30	30/30
	3/30	30/30	30/30
	0/30	26/30	30/30
	0/30	21/30	30/30
	0/30	0/30	24/30

*NP means data not provided in the paper

ACO with parameter adaptation using the proposed methodology, only in the last problem cannot obtain always the minimum and in this last problem on average is better the method from Neyoy et al. [7], but in Table 5.10 are more details about the experimentation with TSP.

From the results in Table 5.10, we want to focus on the fifth problem because on Table 5.9 Neyoy et al. [7] have on average better results in this problem, but results in Table 5.10 shows that our approach can achieve the global minimum several times in this problem, and Neyoy et al. [7] reported that their approach cannot obtain the global minimum of this problem. However, their average on this problem is better than our result; regrettably, in their paper, they do not show their worst results or the standard deviation.

Table 5.10 shows that the proposed approach can achieve the global minimum of the TSP problems in almost every experiment. It can be stated that the proposed methodology for parameter adaptation helps ACO to improve the quality of the results, when compared with the results of the original method, and even when it is compared with approaches that also change in a dynamic way some parameters of ACO.

5.3 Results with the Autonomous Mobile Robot

In the experiments with the optimization of a fuzzy system for controlling an autonomous mobile robot, we consider using the same parameters as in the TSP problems, but only the iterations are different; in this case, only 100 iterations are used. Only because the computing time increases with more iterations, also we found that with only 100 iterations, the methods can find good parameters for the membership functions of the fuzzy controller.

Just to be clear, in the experimentation with the membership functions of a fuzzy system used for controlling an autonomous mobile robot, each method is executed 30 times, so each method produces 30 different fuzzy controllers, in order to make a comparison between them. Also, the methods only try to find the best parameters of the membership functions, so the type of the membership functions is the same, and also the rule set is not changed.

The objective function of the methods is defined as the mean square error (MSE) of the trajectory; we know that there are another metrics for control, so we also show these metrics next to the MSE.

The errors used to measure the performance of the fuzzy controllers are as follows: mean squared error (MSE), integral squared error (ISE), integral absolute error (IAE), integral of time-weighted squared error (ITSE), and integral time-weighted absolute error (ITAE), where they are shown, respectively, in Eqs. (5.1)–(5.5).

$$\text{MSE} = \frac{1}{n} \sum_{i=1}^{n} \left(\hat{Y}_i - Y_i \right)^2 \qquad (5.1)$$

$$\text{ISE} = \int_0^\infty e^2(t) \mathrm{d}t \qquad (5.2)$$

$$\text{IAE} = \int_0^\infty |e(t)| \, \mathrm{d}t \qquad (5.3)$$

$$\text{ITSE} = \int_0^\infty e^2(t) t \mathrm{d}t \qquad (5.4)$$

$$\text{ITAE} = \int_0^\infty |e(t)| \, t \mathrm{d}t \qquad (5.5)$$

The fuzzy system used to control the robot uses the linear and angular errors to decide on the level of torque for each motor of the robot, and these are the errors from Eqs. (5.1) to (5.5) and are based on the linear and angular errors, so in Table 5.11, each of these errors has two results: the first one is based on the linear error and the second one is based on the angular error. In Table 5.11, each result is an average of the 30 experiments; also, the results in bold are the best for each type of error.

An important note from the results in Table 5.11 is that the fuzzy controllers were optimized based on the MSE, and the other errors (based on the other metrics) are calculated for future reference, where some of these errors may have more importance. In this problem, the difference between the trajectory of the robot and the

Table 5.11 Results in the experiments with the autonomous mobile robot

Error	Original ACO	Neyoy et al. [7]	FACO with parameter adaptation
MSE	0.2169	0.0131	0.0096
ISE	1.1729	NP* NP*	1.1081
	2.5021		2.3673
IAE	5.5201	NP* NP*	5.5423
			7.4419
	7.6957		
ITSE	17.6999	NP* NP*	16.7519
	37.1416		34.6626
ITAE	82.8634	NP* NP*	82.4166
	113.5225		108.2381

*NP means data not provided in the paper

Table 5.12 Details from the results in Table 5.11

Error	Original ACO	Neyoy et al. [7]	FACO with parameter adaptation
Best	0.0103	2.9e−4	5.8791e−5
Worst	0.7623	NP*	0.0601
Standard deviation	0.0718	NP*	0.0148

*NP means data not provided in the paper

desired trajectory is more important, and the MSE is the best metric to calculate this difference.

The results in Table 5.11 from the robot problem show that the proposed approach can achieve on average better results when compared with the other methods, and only in the IAE of the linear error, the original algorithm obtains a better result, but with not much of a difference, and in all of the metrics including the MSE, the proposed approach outperforms all of the other methods.

Table 5.12 shows the details from the results of Table 5.11, where these are the best controller, the worst controller, and the standard deviation from all of the 30 experiments of each method. Each result of Table 5.12 is based on the MSE of the reference trajectory and the real trajectory of the robot, created with the corresponding fuzzy controller.

From the results in Table 5.12, we have that the proposed approach can obtain the best of all controllers, and also even in the worst controller, the standard deviation is also lower when compared with all of the other methods. This means that the controllers that the proposed approach optimizes are close together; in other words,

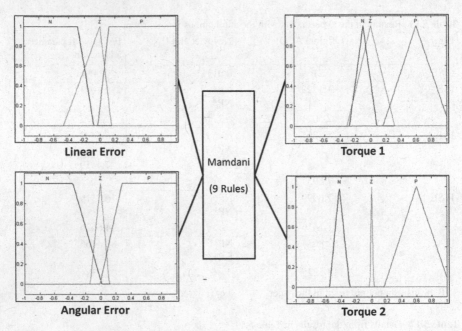

Fig. 5.1 Best fuzzy system used for control of an autonomous mobile robot

the proposed approach can obtain the best controller and the results are good enough with a fine precision, which means that the results are very consistent and accurate.

The best fuzzy controller found by ACO using the proposed methodology for parameter adaptation is illustrated in Fig. 5.1, where we can notice the difference from the fuzzy system used as a basis and illustrated in Fig. 3.5.

With the fuzzy system illustrated in Fig. 5.1 using the fuzzy rule set from Table 3.2, we can obtain the trajectory illustrated in Fig. 5.2. And that trajectory is very different from the trajectory illustrated in Fig. 3.6 created by the fuzzy system used as a basis.

The best fuzzy system used for control of an autonomous mobile robot was created by ACO which uses the proposed methodology for the dynamic adaptation of the parameters α (alpha) and ρ (rho); this fuzzy controller obtains a MSE of $5.8791e-5$.

Also, we want to compare with the trajectory presented by Neyoy et al. [7] and the trajectory presented by Castillo et al. [8], where both consider the same robot problem and the results are illustrated in Fig. 5.3.

Compared with other methodologies like the results in Fig. 5.3, the proposed methodology for parameter adaptation can help ACO to obtain a much better fuzzy controller as appreciated in the trajectories, with only the optimization of the membership functions.

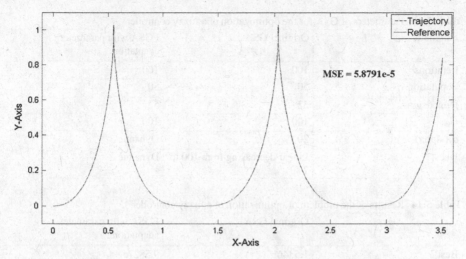

Fig. 5.2 Trajectory of the best fuzzy system used for control of an autonomous mobile robot

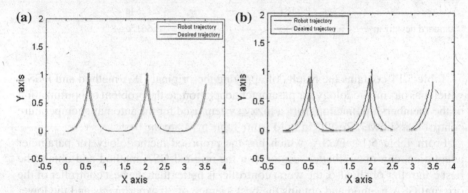

Fig. 5.3 Trajectories of the same robot problem: **a** Neyoy et al. [7] and **b** Castillo et al. [8]

5.4 Results with the Automatic Temperature Control in a Shower

To perform the experimentation with the problem of automatic temperature control in a shower, in this case, GSA is used to optimize de fuzzy controller and use the parameters from Table 5.13.

This problem consists in the optimization of the membership functions from a fuzzy system used to control the level of hot and cold water in a shower. The fuzzy system used as a basis is illustrated in Fig. 3.8 and used the fuzzy rule set from Table 3.3. In this problem, we want to optimize the parameters of the membership functions, so the number of dimensions is 52, which corresponds to the total of points for each membership function.

Table 5.13 Parameters of GSA for the optimization of a fuzzy controller

Parameter	Original GSA	FGSA with parameter adaptation
Iterations	100	100
Population	50	50
Dimensions	52	52
G_0	100	100
α (alpha)	20	Dynamic
Kbest	Linear decreasing from 100 to 2%	Dynamic

Table 5.14 Results of the problem of optimization of a fuzzy controller

	Original GSA	FGSA with parameter adaptation
Best	1.5346	9.5827e−8
Worst	4.1710	4.1371e−7
Average	2.7181	2.1254e−7
Standard deviation	0.7513	5.6615e−8

Table 5.14 contains the results of applying the original GSA method and FGSA which has our methodology for parameter adaptation, to the problem of optimization of the membership function from a fuzzy system used for the automatic temperature control in a shower. Results in bold from Table 5.14 are the best.

From Table 5.14, FGSA, which has the proposed methodology for parameter adaptation, can obtain better results than the original GSA method and obtain the best controller of all; also, the worst controller is better than the best controller of the original GSA method and obtains the best average of 30 experiments and the lower standard deviation.

This benchmark problem of the optimization of a fuzzy controller has been studied by Cervantes et al. [9], where they use a hierarchical genetic algorithm (HGA) and its best experiment was of 9.6e−5, which is a good result but compared with our best experiment which is 9.5827e−8, even our worst experiment is better that the proposed in [9].

The best fuzzy system, which represents the best experiment from Table 5.14, is the fuzzy system illustrated in Fig. 5.4, and its error is 9.5827e−8, this is with the use of the proposed methodology for parameter adaptation GSA can optimize the membership functions from the fuzzy system illustrated in Fig. 3.8 which error is 2.7758.

Figure 5.5 shows the levels of the desired temperature and how the fuzzy controller from Fig. 5.4 tries to follow that desired reference.

Figure 5.6 illustrates the behavior of the fuzzy controller from Fig. 5.4 when it tries to follow the desired flow reference.

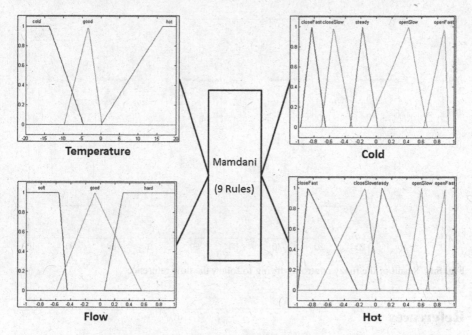

Fig. 5.4 Best fuzzy system used for control the temperature in a shower

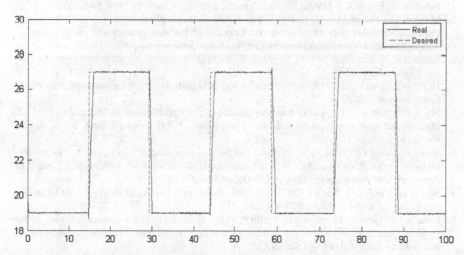

Fig. 5.5 Result of the fuzzy controller trying to follow the temperature reference

GSA using the proposed methodology for parameter adaptation can create a fuzzy controller that obtains an MSE of 9.5827e−8, with only the optimization of all points of the membership functions, and compared with the original GSA method, it is much better in quality of the results.

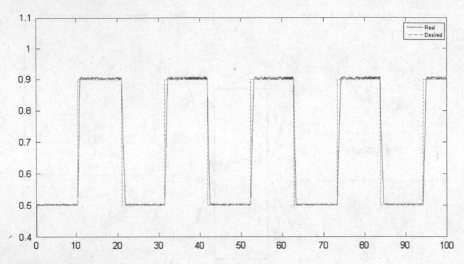

Fig. 5.6 Result of the fuzzy controller trying to follow the flow reference

References

1. Haupt R. L, Haupt S. E (1998) Practical genetic algorithms (Vol. 2). New York, Wiley
2. Marcin M, Smutnicki C (2005) Test functions for optimization needs. http://www.
   bioinformaticslaboratory.nl/twikidata/pub/Education/NBICResearchSchool/Optimization/
   VanKampen/BackgroundInformation/TestFunctions-Optimization.pdf
3. Rashedi E, Nezamabadi-pour H, Saryazdi S (2009) GSA: a gravitational search algorithm. Inf
   Sci 179(13):2232–2248
4. Hongbo L, Ajith A (2007) A fuzzy adaptive turbulent particle swarm optimization. Int J Innov
   Comput Appl 1(1):39–47
5. Shi Y, Eberhart R (2001) Fuzzy adaptive particle swarm optimization. In: Proceeding of IEEE
   international conference on evolutionary computation. IEEE Service Center, Seoul, Korea,
   Piscataway, NJ, pp 101–106
6. Wang B, Liang G, ChanLin W, Yunlong D (2006) A new kind of fuzzy particle swarm optimization fuzzy_PSO algorithm. In: 1st International Symposium on Systems and Control in
   Aerospace and Astronautics, ISSCAA 2006, pp 309–311
7. Neyoy H, Castillo O, Soria J (2012) Dynamic fuzzy logic parameter tuning for ACO and its
   application in TSP Problems. SCI 451:259–271
8. Castillo O, Martinez-Marroquin R, Melin P, Valdez F, Soria J (2012) Comparative study of bio
   inspired algorithms applied to the optimization of type-1 and type-2 fuzzy controllers for an
   autonomous mobile robot. Inf Sci 192:19–38
9. Cervantes L, Castillo O, Melin P (2011) Intelligent control of nonlinear dynamic plants using
   a hierarchical modular approach and type-2 fuzzy logic. In: Mexican international conference
   on artificial intelligence. Springer, Berlin, Heidelberg, pp 1–12
10. Sombra A, Valdez F, Melin P, Castillo O (2013) A new gravitational search algorithm using
    fuzzy logic to parameter adaptation. In: 2013 IEEE Congress on Evolutionary Computation
    (CEC), IEEE, pp 1068–1074

Chapter 6
Statistical Analysis and Comparison of Results

In this chapter is presented a statistical comparison between all the proposed methods with parameter adaptation against the original methods.

Using the statistical test, Z-test with the parameters contained in Table 6.1, and the performance of the proposed methodology for parameter adaptation can be evaluated.

Where μ_1 represents the mean of the results from the proposed method with parameter adaptation using the proposed methodology, and μ_2 represents the mean of the results from the original method without parameter adaptation (which use fixed parameters). With a level of significance of 95% and an alpha of 5%, the rejection region is for all Z values lower than the critical value.

Figure 6.1 illustrates the rejection region of the null hypothesis using the statistical Z-test, for all Z values lower than the critical value $Z_c = -1.645$.

6.1 Comparison of Original PSO Versus Fuzzy PSO

The statistical comparison of the original PSO and the fuzzy PSO with parameter adaptation is made from the results contained in Table 5.1, where each result is an average of 30 experiments, so a Z-test can be performed for each benchmark mathematical function using the parameters contained in Table 6.1.

The results in Table 6.2 are obtained by applying the statistical Z-test to each benchmark mathematical function with the 30 experiments from the function.

In this case, μ_1 is the mean of the results from fuzzy particle swarm optimization with parameter adaptation using the proposed method through an interval type-2 fuzzy system, and μ_2 is the mean of the results from the original particle swarm optimization.

From the results contained in Table 6.2, in all the Z-tests, there are significant evidence to reject the null hypothesis, and this means that the proposed method

© The Author(s) 2018
F. Olivas et al., *Dynamic Parameter Adaptation for Meta-Heuristic Optimization Algorithms Through Type-2 Fuzzy Logic*, SpringerBriefs in Computational Intelligence, https://doi.org/10.1007/978-3-319-70851-5_6

Table 6.1 Parameters for the statistical test Z-test

Parameter	Value
Level of significance	95%
Alpha (α)	5%
Alternative hypothesis (H_a)	$\mu_1 < \mu_2$
Null hypothesis (H_0)	$\mu_1 \geq \mu_2$
Critical value	-1.645

Fig. 6.1 Rejection region of the null hypothesis

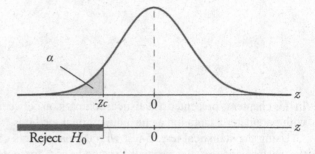

can obtain on average better results than the original method in all the benchmark mathematical functions.

6.2   Comparison of Original ACO Versus Fuzzy ACO

The statistical comparison between the original ACO method against the fuzzy ACO with parameter adaptation using the proposed methodology, is made from the results of the experiments from each TSP problem used to calculate each one of the results contained in Table 5.9.

The results contained in Table 6.3 are the Z-test using the parameters from Table 6.1, in each TSP problem, in this case, μ_1 is the mean of the results of fuzzy ACO with parameter adaptation, and μ_2 is the mean of the results of the original ACO.

From the results in Table 6.3 in all the Z-test, there are significant evidences to reject the null hypothesis, which means that the fuzzy ACO with parameter adaptation using the proposed methodology through an interval type-2 fuzzy system can obtain on average better results than the original ACO in the most complex TSP problems.

From the results with the optimization of the trajectory of an autonomous mobile robot, there are 30 experiments, which are used to calculate the average MSE from Table 5.11 and the standard deviation from Table 5.12. These experiments are used to calculate the Z-test between the original ACO and fuzzy ACO with parameter adaptation in the optimization of membership functions from a fuzzy controller.

Table 6.2 Comparison of results from original PSO against fuzzy PSO using the Z-test

Function	Z value	Evidence
F1	−2.8187	Significant
F2	−3.1446	Significant
F3	−3.0696	Significant
F4	−4.9681	Significant
F5	−2.1006	Significant
F6	−2.0302	Significant
F7	−3.8076	Significant
F8	−4.3258	Significant
F9	−3.8179	Significant
F10	−3.1880	Significant
F11	−2.0344	Significant
F12	−4.5445	Significant
F13	−2.4629	Significant
F14	−3.1068	Significant
F15	−3.9681	Significant
F16	−5.4067	Significant
F17	−6.4399	Significant
F18	−5.8538	Significant
F19	−5.4571	Significant
F20	−3.8451	Significant
F21	−2.0848	Significant
F22	−4.4802	Significant
F23	−3.2606	Significant
F24	−3.5656	Significant
F25	−2.7916	Significant
F26	−2.9762	Significant

Table 6.3 Comparison of results from original ACO against fuzzy ACO using the Z-test

TSP	Z value	Evidence
Burma14	−1.8612	Significant
Ulysses22	−8.3213	Significant
Berlin52	−9.4243	Significant
Eil76	−11.3715	Significant
KroA100	−12.2905	Significant

The result of the Z-test is −5.2944 which is in the rejection region, this means that fuzzy ACO with parameter adaptation using the proposed methodology through an interval type-2 fuzzy system can obtain on average better results than the original ACO in the optimization of membership functions of fuzzy controllers for the trajectory of an autonomous mobile robot.

6.3   Comparison of Original GSA Versus Fuzzy GSA

The statistical comparison between original GSA against fuzzy GSA with parameter adaptation is made from the 30 experiments used to calculate the average MSE and the standard deviation from Table 5.14.

The result of the Z-test is −19.8158, which is in the rejection region, this means that fuzzy gravitational search algorithm with parameter adaptation using the proposed methodology through an interval type-2 fuzzy system can obtain on average better results than the original gravitational search algorithm in the optimization of membership functions from fuzzy controllers for the automatic control of the temperature in a shower.

Chapter 7
Conclusions

During the development of this research work, many papers were published; the proposed methodology for parameter adaptation in meta-heuristic optimization methods shows that it can be a powerful option to add in a new optimization method that uses fixed parameters, because dynamic parameters are better and modeling a desired behavior of the optimization method through fuzzy logic is a great add-on.

The development of the interval type-2 fuzzy system was possible because previously, several type-1 fuzzy systems were created and combined to create a more powerful fuzzy system. Interval type-2 fuzzy logic helps in modeling the fuzzy system because the membership values are themselves fuzzy, which means that a range of membership values can be used instead of only one value.

The results using benchmark mathematical functions were obtained first because of their simplicity, and later changed to TSP, and finally with the optimization of fuzzy controllers, and in all the experiments the proposed methodology helps in finding better results.

From the results of the statistical comparison there is enough evidence to say that the proposed fuzzy PSO with parameter adaptation can obtain on average better results than the original PSO in the benchmark mathematical functions.

From the results with the TSP problems, there is enough evidence in all problems that fuzzy ACO with parameter adaptation using the proposed methodology can obtain on average better results than the original ACO.

From the results in the optimization of a fuzzy controller for an autonomous mobile robot, there is enough evidence that fuzzy ACO with parameter adaptation can obtain on average better results than the original ACO.

From the results with the fuzzy controller for the automatic temperature control in a shower, there is enough evidence that fuzzy GSA with parameter adaptation can obtain on average better results than the original GSA.

The improvement in the optimization algorithm is not only compared with the original method but compared with other improvements to the same optimization algorithm or other methods.

© The Author(s) 2018   51
F. Olivas et al., *Dynamic Parameter Adaptation for Meta-Heuristic Optimization Algorithms Through Type-2 Fuzzy Logic*, SpringerBriefs in Computational Intelligence, https://doi.org/10.1007/978-3-319-70851-5_7

The use of the metrics proposed in this research work can be used with almost all the optimization methods out there, and if anyone wants to use the proposed methodology, they just need to follow the general procedure from Sect. 4.2.

Aside from the meta-heuristic optimization methods used in this research work, there are others who applied our proposed methodology and there was an improvement compared with the original method; some of them are listed in Sect. 1.4 before the list of the papers published during the development of this book.

As future work of this research, it would be the implementation of the proposed methodology on new optimization methods and also could be the creation of new metrics to serve as inputs of the fuzzy system used for parameter adaptation; the metrics used are fine but thinking that better metrics can help in obtaining better results, the proposed methodology can be improved. The use of generalized type-2 fuzzy system can be a good improvement in the proposed methodology. And to use new problems, apply the optimization methods that use the proposed methodology for parameter adaptation.

Appendix

Main fuzzy particle swarm optimization with parameter adaptation through an interval type-2 fuzzy system, applied to the minimization of benchmark mathematical functions
Beginning of file: FPSOCT2forBMF.m

```
%% Fuzzy Particle Swarm Optimization FPSO for benchmark mathematical
function optimization
function [globalpar,globalcost] = FPSOCT2forBMF(Fi)
    % Parameters
    popsize = 30; % Swarm size
    maxit = 100; % Maximum iterations
%      c1=1; c2=3;
    w = 1; % Inertia weight
%      C = 1; % Constriction factor
%      Fi = 4; % TestFunction 1-26
    [npar,LimInf,LimSup] = TestParam(Fi);
    % Read the fuzzy system for parameter adaptation
    fis = readfis('IT2IterationDiversityC1C2Constriction.fis');
    D = zeros(1,maxit+1); % Save diversity each iteration
    vel = rand(popsize,npar); % Random velocities
    par = rand(popsize,npar); % Random population
    cost = zeros(1,popsize);
    for pari=1:popsize
        cost(pari) = TestFunctions(par(pari,:),Fi); % Fitness function
    end
    minc = zeros(1,maxit+1);
    meanc = zeros(1,maxit+1);
    minc(1) = min(cost); % Minimum cost
    meanc(1) = mean(cost); % Mean cost
```

© The Author(s) 2018
F. Olivas et al., *Dynamic Parameter Adaptation for Meta-Heuristic Optimization Algorithms Through Type-2 Fuzzy Logic*, SpringerBriefs in Computational Intelligence https://doi.org/10.1007/978-3-319-70851-5

```
    globalmin = minc(1);
    localpar = par; % Best local minima
    localcost = cost; % Best local cost
    [globalcost,indx] = min(cost); % Search for best particle
    globalpar = par(indx,:); % Best particle
    D(1) = Diversity(par,indx); % Calculate the diversity
    iter = 0; % Iterations counter
    while iter < maxit
        fprintf '.'
        iter = iter + 1;
        r1 = rand(popsize,npar);
        r2 = rand(popsize,npar);
%          w=(maxit-iter)/maxit;
        % Calculate new parameters with the fuzzy system
        it = iter/maxit; % Porcentage of iteration
        out = evalifistype2([it;D(iter)],fis);
        c1 = out(1); c2 = out(2); C = out(3);
        % Velocity update
        vel = C*(w*vel + c1 *r1.*(localpar-par) +
c2*r2.*(ones(popsize,1)*globalpar-par));
        par = par + vel; % Update positions of the particles
        while ~isempty(par(par<LimInf))
            par(par<LimInf) = 1+rand(1)*par(par<LimInf);
        end
        while ~isempty(par(par>LimSup))
            par(par>LimSup) = rand(1)*par(par>LimSup);
        end
        for pari=1:popsize
            cost(pari) = TestFunctions(par(pari,:),Fi); % Fitness function
        end
        bettercost = cost < localcost;
        localcost = localcost.*not(bettercost) + cost.*bettercost;
        localpar(bettercost,:) = par(bettercost,:);
        [temp, t] = min(localcost);
        if temp < globalcost
            globalpar = par(t,:); globalcost = temp;
        end
        [~,pos] = min(cost);
        D(iter+1) = Diversidad(par,pos); % Calculate the diversity
        minc(iter+1) = min(cost); % Minimum of each iteration
        globalmin(iter+1) = globalcost;
        meanc(iter+1) = mean(cost);
    end % End iterations
    disp '\'
```

```
      disp(['FPSOC = F(' num2str(globalpar) ') = ' num2str(globalcost)])
      figure
      iters=0:maxit;
      plot(iters,minc,iters,meanc,'-',iters,globalmin,':');
      xlabel('Iteration');ylabel('Fitness');
      legend('Minimum','Mean','Best')
end
```

End of file: FPSOCT2.m

Interval type-2 fuzzy system used for parameter adaptation in PSO
Beginning of file: IT2IterationDiversityC1C2Constriction.fis

```
[System]
Name='IT2IterationDiversityC1C2Constriction'
Type='mamdani'
Version=2.0
NumInputs=2
NumOutputs=3
NumRules=9
AndMethod='min'
OrMethod='max'
ImpMethod='min'
AggMethod='max'
DefuzzMethod='centroid'

[Input1]
Name='Iteration'
Range=[0 1]
NumMFs=3
MF1='Low':'itritype2',[-0.669457968162042 -0.140620920662161
0.351248786029302 -0.330542031837958
0.140620920662161 0.648751213970698]
MF2='Medium':'itritype2',[-0.177962360710304 0.335883258653366
0.957332537298521 0.177962360710304 0.664116741346634 1.04266746270148]
MF3='high':'itritype2',[0.402920543337201 0.795690706538561
1.34834755908111 0.597079456662799 1.20430929346144 1.65165244091889]

[Input2]
Name='Diversity'
Range=[0 1]
NumMFs=3
MF1='Low':'itritype2',[-0.587718452669538 -0.117983310210837
0.33217224605816 -0.412281547330462 0.117983310210837 0.66782775394184]
MF2='Medium':'itritype2',[-0.101590325993638 0.399965862423552
```

```
0.847077801506029 0.101590325993638 0.600034137576448 1.15292219849397]
MF3='High':'itritype2',[0.333755130945738 0.987021136234523
1.42711946669828 0.666244869054262 1.01297886376548 1.57288053330172]


[Output1]
Name='C1'
Range=[0 3]
NumMFs=5
MF1='Low':'itritype2',[-0.195113099762525 0.369522005755393
0.793941828933252 0.195113099762525 0.630477994244607 1.20605817106675]
MF2='MediumLow':'itritype2',[0.303858720477035 0.793857495691285
1.41668481115705 0.696141279522965 1.20614250430871 1.58331518884295]
MF3='Medium':'itritype2',[0.887858176128502 1.28779061201825
1.91267871896526 1.1121418238715 1.71220938798175 2.08732128103474]
MF4='MediumHigh':'itritype2',[1.3438593720076 1.87266317477356
2.37947469687628 1.6561406279924 2.12733682522644 2.62052530312372]
MF5='High':'itritype2',
[1.94361930123235 2.44276341197114 2.86215399058005
2.05638069876765 2.55723658802886 3.13784600941995]


[Output2]
Name='C2'
Range=[0 3]
NumMFs=5
MF1='Low':'itritype2',[-0.238044868847883 0.353921354336404
0.886264390425851 0.238044868847883 0.646078645663596 1.11373560957415]
MF2='MediumLow':'itritype2',[0.479041025920032 0.838601030629259
1.39549642106665 0.520958974079968 1.16139896937074 1.60450357893335]
MF3='Medium':'itritype2',[0.938853362633504 1.3859269692626
1.94195010556842 1.0611466373665 1.6140730307374 2.05804989443158]
MF4='MediumHigh':'itritype2',[1.39469401048605 1.92046062963969
2.47130417248408 1.60530598951395 2.07953937036031 2.52869582751593]
MF5='High':'itritype2',
[1.93839482051866 2.3793086536781 2.89554459839791
2.06160517948134 2.6206913463219 3.10445540160209]


[Output3]
Name='Constriction'
Range=[0 20]
NumMFs=5
MF1='Low':'itritype2',[-5.33993222420809 -2.25538677229037 3.22385894330498
-4.66006777579191 2.25538677229037 6.77614105669502]
MF2='MediumLow':'itritype2',[-1.96872102314898 4.34644731082391
```

9.37863986056062 1.96872102314898 5.65355268917609 10.6213601394394]
MF3='Medium':'itritype2',[3.20269059848694 9.61758748266701
13.6481791653774 6.79730940151306 10.382412517333 16.3518208346226]
MF4='MediumHigh':'itritype2',[9.10600659332097 13.5901332457941
19.2084197476346 10.893993406679 16.4098667542059 20.7915802523654]
MF5='High':'itritype2',
[14.4463706931555 17.954175867861 22.5552409260849
15.5536293068445 22.045824132139 27.4447590739151]

[Rules]
1 1, 5 1 5 (1) : 1
1 2, 4 2 4 (1) : 1
1 3, 4 2 3 (1) : 1
2 1, 4 2 4 (1) : 1
2 2, 3 3 3 (1) : 1
2 3, 2 4 2 (1) : 1
3 1, 3 5 3 (1) : 1
3 2, 2 4 2 (1) : 1
3 3, 1 5 1 (1) : 1

End of file: IT2IterationDiversityC1C2Constriction.fis

Function used to calculate the diversity of the population
Beginning of file: Diversity.m

```
%% Calculate the diversity of the population
function Diver = Diversity(Population,PosBest)
    [nPart,~] = size(Population);
    Best = Population(PosBest,:);
    Population(PosBest,:) = [];
    D1 = zeros(nPart-1,1);
    for i=1:nPart-1
        D1(i) = sqrt(sum((Population(i,:)-Best).^2));
    end
    dis = (1/nPart)*sum(D1);
    if min(D1)==max(D1);
        Diver=0;
    else
        Diver = (dis-min(D1))/(max(D1)-min(D1)); % Normalize
    end
    if Diver>1; Diver=1; end;
    if Diver<0; Diver=0; end;
end
```

End of file: Diversity.m

Function which contains all the benchmark mathematical functions used as fitness function
Beginning of file: TestFunctions.m

```
%% Test functions for optimization
function f=TestFunctions(x,funnum)
    if funnum==1 %F1 Dim=1 Lim=-20:20 Min=f(0)=1
        f=abs(x(1))+cos(x(1));
    elseif funnum==2 %F2 Dim=1 Lim=-20:20 Min=f(0)=0
        f=abs(x(1))+sin(x(1));
    elseif funnum==3 %1st DeJong's Dim=N Lim=-5.12:5.12 Min=f(0,0)=0
        f(:,1)=sum(x'.^2);
    elseif funnum==4 %Rosenbrock Dim=N Lim=-2.048:2.048 Min=f(1,1)=0
        [~,Dim]=size(x); f=0;
        for j=1:Dim-1;
            f=f+100*(x(j+1)^2-x(j))^2+(1-x(j))^2;
        end
    elseif funnum==5 %F5 Dim=N Lim=-10:10 Min=f(0,0)=-20
        f(:,1)=sum(abs(x')-10*cos(sqrt(abs(10*x'))))';
    elseif funnum==6 %F6 Dim=1 Lim=-10:10 Min=f(9.6204)=-100.2238
        f=(x(1).^2+x(1)).*cos(x(1));
    elseif funnum==7 %F7 Dim=2 Lim=0:10 Min=f(9.039,8.668)=-18.5547
        f=x(:,1).*sin(4*x(:,1))+1.1*x(:,2).*sin(2*x(:,2));
    elseif funnum==8 %Axis parallel hyper-ellipsoid function Dim=2: Lim=-
5.12:5.12 Min=f(0,0)=0
        f=x(:,1).^2+(2*x(:,2).^2);
    elseif funnum==9 %Griewangk Dim=N Lim=-600:600 Min=f(0,0)=0
        [~,Dim]=size(x); fr=4000; s=0; p=1;
        for j=1:Dim; s=s+x(j)^2; end
        for j=1:Dim; p=p*cos(x(j)/sqrt(j)); end
        f = s/fr-p+1;
    elseif funnum==10 %F12 Dim=2 Lim=-5:5 Min=f(1.897,1.006)=-0.5231
        f(:,1)=.5+(sin(sqrt(x(:,1).^2+x(:,2).^2).^2)-
.5)./(1+.1*(x(:,1).^2+x(:,2).^2));
    elseif funnum==11 %F13 Dim=2 Lim=-10:10 Min=f(0,0)=0
        aa=x(:,1).^2+x(:,2).^2;
        bb=((x(:,1)+.5).^2+x(:,2).^2).^0.1;
        f(:,1)=aa.^0.25.*sin(30*bb).^2+abs(x(:,1))+abs(x(:,2));
    elseif funnum==12 %F14 Dim=2 Lim=-5:5 Min=f(1,1.6606)=-0.3356
        f(:,1)=besselj(0,x(:,1).^2+x(:,2).^2)+abs(1- x(:,1))/10+abs(1-
x(:,2))/10;
    elseif funnum==13 %F15 Dim=2 Lim=-5:5 Min=f(-2.7730,-5)=-16.9473
        f(:,1)=-exp(.2*sqrt((x(:,1)-1).^2+(x(:,2)-
1).^2)+(cos(2*x(:,1))+sin(2*x(:,1))));
    elseif funnum==14 %F16 Dim=2 Lim=-20:20 Min=f(-14.58,-20)=-23.8062
```

```
        f(:,1)=x(:,1).*sin(sqrt(abs(x(:,1)-(x(:,2)+9))))-(x(:,2)+9).*...
            sin(sqrt(abs(x(:,2)+0.5*x(:,1)+9)));
    elseif funnum==15 %Schwefel Dim=2 Lim=-500:500
Min=f(420.9687,420.9687)=-837.9658
        [~,Dim]=size(x);
        s=sum(-x.*sin(sqrt(abs(x))));
        f=418.9829*Dim+s;
    elseif funnum==16 %Rotated hyper-ellipsoid function Dim=2 Lim=-
65.536:65.536 Min=f(0,0)=0
        f(:,1)=x(:,1).^2+(x(:,1).^2+x(:,2).^2);
    elseif funnum==17 %Sum of different power functions Dim=2 Lim=-1:1
Min=f(0,0)=0
        f(:,1)=abs(x(:,1)).^2+abs(x(:,2)).^3;
    elseif funnum==18 %Ackley Dim=2 Lim=-32.768:32.768 Min=f(0,0)=0
(8.8818e-016)
        [~,Dim]=size(x); a=20; b=0.2; c=2*pi; s1=0; s2=0;
        for i=1:Dim; s1=s1+x(i)^2; s2=s2+cos(c*x(i)); end;
        f=-a*exp(-b*sqrt(1/Dim*s1))-exp(1/Dim*s2)+a+exp(1);
    elseif funnum==19 % Michalewicz Dim=2 Lim=0:pi Min=f(2.2030,1.5710)=-
1.8013
        f(:,1)=-(sin(x(:,1)).*((sin((x(:,1).^2)/pi)).^20))-...
            (sin(x(:,2)).*((sin((2^x(:,2).^2)/pi)).^20));
    elseif funnum==20 % Branins Dim=2 Lim=-4:14 Min=f((-
pi,12.275);(pi,2.275);(9.42478,2.475))=3.9270
        a=1;b=(5.1/(4*pi^2));c=5/pi;d=6;e=10;t=1/8*pi;
        f=a*((x(:,2)-(b*x(:,1).^2)+c*x(:,1)-d).^2)+e*(1-f)*cos(x(:,1))+e;
    elseif funnum==21 % Easom Dim=2 Lim=-100:100 Min=f(pi,pi)=-1
        f(:,1)=-cos(x(:,1)).*cos(x(:,2)).*exp(-((x(:,1)-pi).^2)-((x(:,2)-
pi).^2));
    elseif funnum==22 % Goldstein-Price Dim=2 Lim=-2:2 Min=f(0,-1)=3
        f(:,1)=(1+((x(:,1)+x(:,2)+1).^2).*(19-(14*x(:,1))+(3*(x(:,1).^2))-
...
(14*x(:,2))+(6*x(:,1).*x(:,2))+(3*(x(:,2).^2)))).*(30+(((2*x(:,1))-...
            (3*x(:,2))).^2).*(18-
(32*x(:,1))+(12*(x(:,1).^2))+(48*x(:,2))-...
            (36*x(:,1).*x(:,2))+(27*(x(:,2).^2)))));
    elseif funnum==23 % Six-hump camel back function Dim=2 Lim=-2:2
Min=f((-0.0898,0.7126);(0.0898,-0.7126))=-1.0316
        f(:,1)=((4-
(2.1*(x(:,1).^2))+((x(:,1).^4)/3)).*(x(:,1).^2))+(x(:,1).*...
            x(:,2))+((-4+(4*(x(:,2).^2))).*(x(:,2).^2));
    elseif funnum==24 % Fifth function of DeJong Dim=2 Lim=-65.536:65.536
Min=f(-32,-32)=0.9980
```

```
        a=[-32,-16,0,16,32,-32,-16,0,16,32,-32,-16,0,16,32,-32,-
16,0,16,32,-32,-16,0,16,32;
            -32,-32,-32,-32,-32,-16,-16,-16,-16,-
16,0,0,0,0,0,16,16,16,16,16,32,32,32,32,32];
        f=zeros(size(x,1),1);
        for i=1:size(x,1)
            p=x(i,:); k=0.002;
            for j=1:25; k=k+1/(j+(p(1)-a(1,j))^6+(p(2)-a(2,j))^6); end;
            f(i)=1/k;
        end
    elseif funnum==25 %Drop wave function Dim=2 Lim=-5.12:5.12 Min=f(0,0)=-
1
        f(:,1)=-
((1+cos(12*sqrt((x(:,1).^2)+(x(:,2).^2))))./((1/2*((x(:,1).^2)+(x(:,2).^2))
)+2));
    elseif funnum==26 % Shubert Dim=2 Lim=-5.12:5.12 Min=f(-0.2,-0.2)=-
200.4818
        s1=0; s2=0;
        for i=1:5
            s1=s1+i*cos((i+1)*x(1)+i);
            s2=s2+i*cos((i+1)*x(2)+i);
        end
        f=s1*s2;
    elseif funnum==27 % Rastrigin Dim=N Lim=-5.12:5.12 Min=f(0,0)=0
        [~,Dim]=size(x); s=0;
        for j=1:Dim; s=s+(x(j)^2-10*cos(2*pi*x(j))); end;
        f=10*Dim+s;
    end
end
```

End of file: TestFunctions.m

Function which contains the limits of the search space from all the benchmark mathematical functions
Beginning of file: _TestParam.m_

```
%% Limits of the Test functions for optimization
function [npar,LimI,LimS] = TestParam(Fi)
    % Limits
    Lim = [-20,20; -20,20; -5.12,5.12; -2.048,2.048; -10,10;...1-5
            -10,10; 0,10; -5.12,5.12; -600,600;...6-9
            -5,5; -10,10; -5,5; -5,5; -20,20; -500,500;...10-15
            -65.536,65.536; -1,1; -32.768,32.768; 0,pi;...16-19
            4,14; -100,100; -2,2; -2,2; -65.536,65.536;...20-24
```

```
               -5.12,5.12; -5.12,5.12; -5.12,5.12;]; %25-27
    % Outputs
    npar = 30;
    LimI = Lim(Fi,1);
    LimS = Lim(Fi,2);
end
```

End of file: TestParam.m

Main fuzzy particle swarm optimization with parameter adaptation through an interval type-2 fuzzy system, applied to the optimization of membership functions from fuzzy controllers
Beginning of file: FPSOCT2forMF.m

```
%% Fuzzy Particle Swarm Optimization FPSO for fuzzy system optimization
function [globalpar,globalcost] = FPSOCT2forMF()
    % Parameters
    popsize = 30;  % Swarm size
    maxit = 100;    % Maximum iterations
    w = 1; % Inercia weight
    % Read the fuzzy system for parameter adaptation
    FIS = readfis('IT2IterationDiversityC1C2Constriction.fis');
    FisBase = readfis('BASE.fis'); % Basic fuzzy controller
    D = zeros(1,maxit+1);
    npar = nPar(FisBase); % Dimensions
    par = rand(popsize,npar); % Random population
    par = ValidatePopulation(FisBase,par); % Validate Population
    vel = rand(popsize,npar); % Random velocities
    cost = FitnessFunction(FisBase,par);
    minc = zeros(1,maxit+1);
    meanc = zeros(1,maxit+1);
    minc(1) = min(cost);
    meanc(1) = mean(cost);
    globalmin = minc(1);
    localpar = par;
    localcost = cost;
    [globalcost,indx] = min(cost);
    globalpar = par(indx,:);
    D(1) = Diversity(par,indx);
    iter = 0;
    while iter < maxit
        disp(num2str(iter))
        iter = iter + 1;
        r1 = rand(popsize,npar);
        r2 = rand(popsize,npar);
        % Calculate the new parameters with the fuzzy system
```

```
           it = iter/maxit;
           out = evalifistype2([it;D(iter)],FIS);
           c1 = out(1); c2 = out(2); C = out(3);
           % Update velocity
           vel = C*(w*vel + c1 *r1.*(localpar-par) +
c2*r2.*(ones(popsize,1)*globalpar-par));
           par = par + vel; % Update positions
           par = ValidatePopulation(FisBase,par); % Validate population
           cost = FitnessFunction(FisBase,par);
           bettercost = cost < localcost;
           localcost = localcost.*not(bettercost) + cost.*bettercost;
           localpar(bettercost,:) = par(bettercost,:);
           [temp, t] = min(localcost);
           if temp < globalcost
               globalpar = par(t,:); globalcost = temp;
           end
           [~,pos] = min(cost);
           D(iter+1) = Diversity(par,pos);
           minc(iter+1) = min(cost);
           globalmin(iter+1) = globalcost;
           meanc(iter+1) = mean(cost);
       end % End iterations
       disp(num2str(iter))
end
```

End of file: FPSOCT2forMF.m

Fitness function used for optimization of membership functions
Beginning of file: FitnessFunction.m

```
%% Fitness Function for optimization of MFs
% FIS = Is the basis fuzzy system to optimize
% Population = Is the population
function Fitness = FitnessFunction(FIS,Population)
    [N,y] = size(Population);
    Fitness = zeros(1,N);
    for i=1:N
        fprintf '.'
        Particle = Population(i,:);
        FIS = Particle2FIS(FIS,Particle); % Convert a particle into a FIS
        assignin('base','fis',FIS);
        sim('PRobot1'); % Simulation
        LinearMSE = mean((data(:,2) - data(:,1)).^2); % Lineal Error
        AngularMSE = mean((data(:,4) - data(:,3)).^2); % Angular Error
        Fitness(i) = LinearMSE + AngularMSE; % Fitness of the Particle
```

```
        end
end
```

End of file: FitnessFunction.m

Function used to convert a particle into a fuzzy inference system
Beginning of file: Particle2FIS.m

```
%% Convert a particle into a FIS
% The particle needs to be evaluated before
% FIS = Basis fuzzy system
% Particle = Is a particle
function FIS = Particle2FIS(FIS,Particle)
    cont = 1;
    for i=1:size(FIS.input,2) % For each input
        for j=1:size(FIS.input(i).mf,2) % For each MFs
            if strcmp(FIS.input(i).mf(j).type,'trimf') % Triangular
                FIS.input(i).mf(j).params = Particle(cont:cont+2);
                cont=cont+3;
            elseif strcmp(FIS.input(i).mf(j).type,'trapmf') % Trapezoidal
                FIS.input(i).mf(j).params = Particle(cont:cont+3);
                cont=cont+4;
            end
        end
    end
    for i=1:size(FIS.output,2) % For each input
        for j=1:size(FIS.output(i).mf,2) % For each MFs
            if strcmp(FIS.output(i).mf(j).type,'trimf') % Triangular
                FIS.output(i).mf(j).params = Particle(cont:cont+2);
                cont=cont+3;
            elseif strcmp(FIS.output(i).mf(j).type,'trapmf') % Trapezoidal
                FIS.output(i).mf(j).params = Particle(cont:cont+3);
                cont=cont+4;
            end
        end
    end
end
```

End of file: Particle2FIS.m

Function used to validate the parameter contained in a particle which will be converted into a fuzzy inference system
Beginning of file: ValidateParticle.m

```
%% Validate a particle to be converted into a FIS
% FIS = Basis fuzzy system
% Particle = Has the parameters to be evaluated
```

```
function Particle = ValidateParticle(FIS,Particle)
    cont=0;
    for i=1:size(FIS.input,2) % For each input
        for j=1:size(FIS.input(i).mf,2) % For each MFs
            if strcmp(FIS.input(i).mf(j).type,'trimf') % Triangular

x=sort([Particle(cont+1),Particle(cont+2),Particle(cont+3)]);
                Particle(cont+1) = x(1);
                Particle(cont+2) = x(2);
                Particle(cont+3) = x(3);
                cont=cont+3;
            elseif strcmp(FIS.input(i).mf(j).type,'trapmf') % Trapezoidal

x=sort([Particle(cont+1),Particle(cont+2),Particle(cont+3),Particle(cont+4)
]);
                Particle(cont+1) = x(1);
                Particle(cont+2) = x(2);
                Particle(cont+3) = x(3);
                Particle(cont+4) = x(4);
                cont=cont+4;
            end
        end
    end
    for i=1:size(FIS.output,2) % For each Output
        out = [1,2,4,3,5,7,6,8,9];
        x = Particle(cont+1:cont+9);
        x = sort(x);
        for j=1:9
            z=out(j);
            Particle(cont+j) = x(z);
        end
        cont = cont+9;
    end
end
```

End of file: ValidateParticle.m

Function used to validate the population, to bring the particles back to the search space and not get lost
Beginning of file: ValidatePopulation.m

```
%% Validate the population
% FIS = Basis fuzzy system
% Pop = Population
function Pop = ValidatePopulation(FIS,Pop)
```

```
Lim = [-2,-1, -2,-1, -1,0, -1,0,... % Input1 Trap
       -1,0, -0.1,0.1, 0,1,... % Input1 Tri
       0,1, 0,1, 1,2, 1,2,... % Input1 Trap
       -2,-1, -2,-1, -1,0, -1,0,... % Input2 Trap
       -1,0, -0.1,0.1, 0,1,... % Input2 Tri
       0,1, 0,1, 1,2, 1,2,... % Input2 Trap
       -1,1, -1,1, -1,1,... %-2,-1, -1,0, -1,0,... % Output1 Tri
       -1,1, -1,1, -1,1,... %-1,0, -0.1,0.1, 0,1,... % Output1 Tri
       -1,1, -1,1, -1,1,... %0,1, 0,1, 1,2,... % Output1 Tri
       -1,1, -1,1, -1,1,... %-2,-1, -1,0, -1,0,... % Output2 Tri
       -1,1, -1,1, -1,1,... %-1,0, -0.1,0.1, 0,1,... % Output2 Tri
       -1,1, -1,1, -1,1]; %0,1, 0,1, 1,2]; % Output2 Tri
[x,y] = size(Pop);
for i=1:x % Number of particles
    k=1;
    for j=1:y % Number of points of the MFs
        if Pop(i,j) < Lim(k) || Pop(i,j) > Lim(k+1)
            Pop(i,j) = rand(1)*(Lim(k+1)-Lim(k))+Lim(k);
        end
        k=k+2;
    end
    Pop(i,:) = ValidateParticle(FIS,Pop(i,:));
end
end
```

End of file: ValidatePopulation.m

Main Fuzzy Ant Colony Optimization with parameter adaptation through an interval type-2 fuzzy system, applied to the minimization of the Travel Salesman Problems

Beginning of file: FACOT2forTSP.m

```
%% Fuzzy Ant Colony Optimization for TSP problems
function [bs_tour,bs_tlen] = FACOT2forTSP()
    % TSP problems
    infile='.\TSP\burma14.tsp';
%     infile='.\TSP\ulysses22.tsp';
%     infile='.\TSP\berlin52.tsp';
%     infile='.\TSP\eil76.tsp';
%     infile='.\TSP\kroA100.tsp';
    % Parameters
    maxit=500; % Maximum iterations
    m=0; % Number of ants
    nn_length=0; % Size of the nearest neighborhood
```

```matlab
    alpha=1; beta=2; rho=0.5; w=6; e=0;
    [~,~,n,m,dist,nn_list,choice_info,pheromone,ant] =
InitializeData(infile,nn_length,m,alpha,beta,rho,e,w);
    bsf_ant =
struct('tour_length',Inf,'tour',zeros(1,n),'visited',zeros(1,n));
    bs_tour = [];
    lengths = zeros(maxit,3);
    branching_avg = zeros(maxit,1);
    % Read the fuzzy system for parameter adaptation
    fis = readfis('IT2IterationDiversityAlphaRho.fis');
    for i = 1:maxit
        ant =
ConstructSolutions(n,m,nn_list,dist,choice_info,ant,nn_length);
        [bs_tour,lengths,branching_avg] =
UpdateStatistics(bsf_ant,bs_tour,lengths,ant,i,pheromone,n,branching_avg,di
st);
        [bi_tlen,index] = min([ant(:).tour_length]);
        if(bi_tlen < bsf_ant.tour_length)
            bsf_ant = ant(index);
        end
        bs_tlen = bsf_ant.tour_length;
        % Calculate the new parameters using a fuzzy system
        Iter = i/maxit;
        [~,indx] = min([ant(:).tour_length]);
        ant2 = Ant2Mat(ant);
        Diver = Diversity(ant2,indx);
        out = evalifistype2([Iter;Diver],fis);
        alpha = out(1); rho = out(2);
        [pheromone,choice_info] =
ASRankPheromoneUpdate(w,ant,bsf_ant,n,rho,pheromone,dist,alpha,beta,Q);
    end % End iterations
end


%% Function to initialize the data, read the instance of the TSP problem
function [xcity,ycity,n,m,dist,nn_list,choice_info,pheromone,ant] =
InitializeData(infile,nn_length,m,alpha,beta,rho,w)
    [n,node_coord,~,~,weightType] = ReadInstance(infile);
    xcity = node_coord(:,2);
    ycity = node_coord(:,3);
    dist = zeros(n,n);
    if strcmp(weightType,char('GEO'))
        [latitude,longitude] = GEOtoLATLON(xcity,ycity);
        for i = 1:n
```

```
            for j = 1:i
                dist(i,j) =
GEODistance(latitude(i),longitude(i),latitude(j),longitude(j));
                dist(j,i) = dist(i,j);
            end
        end
    elseif strcmp(weightType,'EUC_2D')
        for i = 1:n
            for j = 1:i
                dist(i,j) = round(sqrt((xcity(i)-xcity(j))^2+(ycity(i)-
ycity(j))^2));
                dist(j,i) = dist(i,j);
            end
        end
    else
        err = MException('ResultChk:BadInput','TSP Node weight type not
supported');
            throw(err)
    end
    [~,nn_list] = sort(dist,2,'ascend');
    nn_list = nn_list(:,1:nn_length);
    if (m == 0)
        m = n;
    end
    ant(m) =
struct('tour_length',0,'tour',randperm(n),'visited',zeros(1,n));
    nnt_length = NearestNeighborTourLength(ant(m),n,dist);
    r = w - 1;
    pheromone = ((0.5*r*(r-1))/(rho*nnt_length))*ones(n,n);
    for i = 1:n
        pheromone(i,i) = 0;
    end
    choice_info = ComputeChoiceInformation(n,pheromone,dist,alpha,beta);
end


%% Function to read an instance of a TSP
function [Dimension,NodeCoord,NodeWeight,Name,WeightType] =
ReadInstance(infile)
    cd TSP
    if ischar(infile)
        fid=fopen(infile,'r');
    else
        disp('input file no exist');
        return;
    end
```

```
        if fid<0
            disp('error while open file');
            return;
        end
        cd ..
        NodeWeight = [];
        while feof(fid)==0
            temps=fgetl(fid);
            if strcmp(temps,'')
                continue;
            elseif strncmpi('NAME',temps,4)
                k=findstr(temps,':');
                Name=temps(k+1:length(temps));
            elseif strncmpi('DIMENSION',temps,9)
                k=findstr(temps,':');
                d=temps(k+1:length(temps));
                Dimension=str2double(d); %str2numz
            elseif strncmpi('EDGE_WEIGHT_TYPE',temps,16)
                k=findstr(temps,':');
                WeightType=temps(k+2:length(temps));
            elseif strncmpi('EDGE_WEIGHT_SECTION',temps,19)
                formatstr = [];
                for i=1:Dimension
                    formatstr = [formatstr,'%g '];
                end
                NodeWeight=fscanf(fid,formatstr,[Dimension,Dimension]);
                NodeWeight=NodeWeight';
            elseif strncmpi('NODE_COORD_SECTION',temps,18) ||
strncmpi('DISPLAY_DATA_SECTION',temps,20)
                NodeCoord=fscanf(fid,'%g %g %g',[3 Dimension]);
                NodeCoord=NodeCoord';
            end
        end
        fclose(fid);
end


%% Function to convert GEO into latitude and longitude
function [latitude,longitude] = GEOtoLATLON(x,y)
        deg = fix(x);
        min = x - deg;
        latitude = pi * (deg+5.0*min/3.0)/180.0;
        deg = fix(y);
```

```
    min = y - deg;
    longitude = pi *(deg+5.0*min/3.0)/180.0;
end


%% Function to calculate the GEO distance
function distance = GEODistance(latitudei,longitudei,latitudej,longitudej)
    RRR = 6378.388;
    q1 = cos(longitudei-longitudej);
    q2 = cos(latitudei-latitudej);
    q3 = cos(latitudei+latitudej);
    distance = fix((RRR*acos(0.5*((1.0+q1)*q2 - (1.0-q1)*q3))+1.0));
end


%% Function to calculate the nearest neighbor tour length
function length = NearestNeighborTourLength(ant,n,dist)
    step = 1;
    r = randi(n,1);
    ant.tour(step) = r;
    ant.visited(r) = 1;
    while step < n
        step = step + 1;
        v = Inf;
        c = ant.tour(step-1);
        for i = 1:n
          for j = 1:n
              if ~ant.visited(j)
                  if dist(c,j) <= v
                      nc = j;
                      v = dist(c,j);
                  end
              end
          end
        end
        ant.tour(step) = nc;
        ant.visited(nc) = 1;
    end
    ant.tour(n+1) = ant.tour(1);
    ant.tour_length = ComputeTourLength(dist,ant);
    length = ant.tour_length;
end


%% Function to calculate the tour length of an ant
function length = ComputeTourLength(dist,ant)
    length = 0;
```

```
        for i = 1:size(ant.tour,2)-1
            length = length + dist(ant.tour(i),ant.tour(i+1));
        end
end


%% Function to calculate the choice information
function choice_info =
ComputeChoiceInformation(n,pheromone,dist,alpha,beta)
        choice_info = zeros(n,n);
        for i = 1:n
            for j = 1:i
                choice_info(i,j) =
(pheromone(i,j)^alpha)*((1/(dist(i,j)+0.1))^beta);
                choice_info(j,i) = choice_info(i,j);
            end
        end
end


%% Function to construct the tours of the ants
function ant =
ConstructSolutions(n,m,nn_list,dist,choice_info,ant,nn_length)
        for k = 1: m
            ant(k).visited = zeros(1,n);
        end
        step = 1;
        for k = 1:m
            r = randi(n,1);
            ant(k).tour(step) = r;
            ant(k).visited(r) = 1;
        end
        if(nn_length > 0)
            while step < n
                step = step + 1;
                for k = 1:m
                    ant(k) =
NeighborListASDecisionRule(n,nn_list,ant(k),choice_info,step);
                end
            end
        else
            while step < n
            step = step + 1;
            for k = 1:m
```

```
                    ant(k) = ASDecisionRule(n,ant(k),choice_info,step);
            end
            end
        end
        for k = 1:m
            ant(k).tour(n+1) = ant(k).tour(1);
            ant(k).tour_length = ComputeTourLength(dist,ant(k));
        end
end


%% Function to calculate the probability of choice
function ant = ASDecisionRule(n,ant,choice_info,i)
    c = ant.tour(i-1);
    sum_probabilities = 0.0;
    selection_probability = zeros(1,n);
    for j = 1:n
        if ant.visited(j)
            selection_probability(j) = 0.0;
        else
            selection_probability(j) = choice_info(c,j);
            sum_probabilities = sum_probabilities +
selection_probability(j);
        end
    end
    r = sum_probabilities*rand(1);
    j = 1;
    p = selection_probability(j);
    while p <= r && j<n
        j = j + 1;
        p = p + selection_probability(j);
    end
    ant.tour(i) = j;
    ant.visited(j) = 1;
end


%% Function to calculate the neighbor list AS decision rule
function ant = NeighborListASDecisionRule(n,nn_list,ant,choice_info,i)
    nn = size(nn_list,2);
    c = ant.tour(i-1);
    sum_probabilities = 0.0;
    selection_probability = zeros(1,nn);
    for j = 1:nn
        if ant.visited(nn_list(c,j))
            selection_probability(j) = 0.0;
```

```
            else
                selection_probability(j) = choice_info(c,nn_list(c,j));
                sum_probabilities = sum_probabilities +
selection_probability(j);
            end
        end
        if(sum_probabilities == 0)
            ant = ChooseBestNext(ant,n,choice_info,i);
        else
            r = sum_probabilities*rand(1);
            j = 1;
            p = selection_probability(j);
            while p < r
                j = j + 1;
                p = p + selection_probability(j);
            end
            ant.tour(i) = nn_list(c,j);
            ant.visited(nn_list(c,j)) = 1;
        end
end


%% Function to calculate the choose best next
function ant = ChooseBestNext(ant,n,choice_info,i)
    v = 0.0;
    c = ant.tour(i-1);
    for j = 1:n
        if ~ant.visited(j)
            if choice_info(c,j) >= v
                nc = j;
                v = choice_info(c,j);
            end
        end
    end
    ant.tour(i) = nc;
    ant.visited(nc) = 1;
end


%% Function to update the statistics
function [bs_tour,lengths,branching_avg] =
UpdateStatistics(bsf_ant,bs_tour,lengths,ant,i,pheromone,n,branching_avg,di
st)
    [bi_tlen,index] = min([ant(:).tour_length]);
    bs_tlen = bsf_ant.tour_length;
    if(bi_tlen < bs_tlen)
```

```
            bs_tlen = bi_tlen;
            bs_tour = ant(index).tour;
        end
        m_tlen = mean([ant(:).tour_length]);
        lengths(i,:) = [m_tlen bi_tlen bs_tlen];
        branching = node_branching(0.05,pheromone,n,dist);
        branching_avg(i,1) = branching;
end


%% Function to calculate the node branching
function branching = node_branching(l,pheromone,n,dist)
    [~,nn_list] = sort(dist,2,'ascend');
    nn_length = n-1;
    num_branches = zeros(n,1);
    for m = 1:n
        minimum = pheromone(m,nn_list(m,2));
        maximum =  pheromone(m,nn_list(m,2));
        for i = 2:nn_length
            if pheromone(m,nn_list(m,i)) > maximum
                maximum = pheromone(m,nn_list(m,i));
            end
            if pheromone(m,nn_list(m,i)) < minimum
                minimum = pheromone(m,nn_list(m,i));
            end
        end
        cutoff = minimum + l * (maximum - minimum);
        for i = 1:nn_length
            if pheromone(m,nn_list(m,i)) > cutoff
                num_branches(m) = num_branches(m) + 1;
            end
        end
    end
    branching = mean(num_branches);
end


%% Function to convert an ant into a matrix
function ants = Ant2Mat(ant)
    [~,n] = size(ant);
    [~,Dim] = size(ant(1).tour);
    ants=zeros(n,Dim);
    for i=1:n
        ants(i,:) = ant(i).tour;
    end
```

```
end


%% Function to update the pheromone in AS rank-based
function [pheromone,choice_info] =
ASRankPheromoneUpdate(w,ant,bsf_ant,n,rho,pheromone,dist,alpha,beta,Q)
      pheromone = Evaporate(n,rho,pheromone);
      [~,index] = sort([ant(:).tour_length]);
      ant = ant(index);
      pheromone = ASRankBSAntDepositPheromone(w,n,bsf_ant,pheromone,Q);
      for k = 1:w-1
          pheromone = ASRankDepositPheromone(w,k,n,ant(k),pheromone,Q);
      end
      choice_info = ComputeChoiceInformation(n,pheromone,dist,alpha,beta);
end


%% Function to evaporate the pheromone from all the arcs
function pheromone = Evaporate(n,rho,pheromone)
      for i = 1:n
          for j = 1:n
              pheromone(i,j) = (1-rho)*pheromone(i,j);
          end
      end
end


%% Function to deposite the pheromone of the best ant in the arcs
function pheromone = ASRankBSAntDepositPheromone(w,n,bs_ant,pheromone,Q)
      phe_inc = w*(Q/bs_ant.tour_length);
      for i = 1:n
          j = bs_ant.tour(i);
          l = bs_ant.tour(i + 1);
          pheromone(j,l) = pheromone(j,l) + phe_inc;
          pheromone(l,j) = pheromone(j,l);
      end
end


%% Function to deposit the pheromone from all the ants in the arcs
function pheromone = ASRankDepositPheromone(w,r,n,ant,pheromone,Q)
      phe_inc = (w-r)*(Q/ant.tour_length);
      for i = 1:n
          j = ant.tour(i);
          l = ant.tour(i + 1);
          pheromone(j,l) = pheromone(j,l) + phe_inc;
```

```
            pheromone(1,j) = pheromone(j,1);
        end
end
```

End of file: FACOT2forTSP.m

Interval type-2 fuzzy system used for parameter adaptation in ACO
Beginning of file: IT2IterationDiversityAlphaRho.fis

```
[System]
Name='IT2IterationDiversityAlphaRho'
Type='mamdani'
Version=2.0
NumInputs=2
NumOutputs=2
NumRules=9
AndMethod='min'
OrMethod='max'
ImpMethod='min'
AggMethod='max'
DefuzzMethod='centroid'

[Input1]
Name='Iteration'
Range=[0 1]
NumMFs=3
MF1='Low':'itritype2',[-0.59 -0.09 0.41 -0.41 0.09 0.59]
MF2='Medium':'itritype2',[-0.09 0.41 0.91 0.09 0.59 1.09]
MF3='High':'itritype2',[0.41 0.91 1.41 0.59 1.09 1.59]

[Input2]
Name='Diversity'
Range=[0 1]
NumMFs=3
MF1='Low':'itritype2',[-0.59 -0.09 0.41 -0.41 0.09 0.59]
MF2='Medium':'itritype2',[-0.09 0.41 0.91 0.09 0.59 1.09]
MF3='High':'itritype2',[0.41 0.91 1.41 0.59 1.09 1.59]

[Output1]
Name='Alpha'
Range=[0 1]
NumMFs=5
MF1='Low':'itritype2',[-0.05 0.1167 0.2833 0.05 0.2167 0.3833]
MF2='MediumLow':'itritype2',[0.1167 0.2833 0.45 0.2167 0.3833 0.55]
MF3='Medium':'itritype2',[0.2833 0.45 0.6167 0.3833 0.55 0.7167]
```

```
MF4='MediumHigh':'itritype2',[0.45 0.6167 0.7833 0.55 0.7167 0.8833]
MF5='High':'itritype2',[0.6167 0.7833 0.95 0.7167 0.8833 1.05]

[Output2]
Name='Rho'
Range=[0 1]
NumMFs=5
MF1='Low':'itritype2',[-0.05 0.1167 0.2833 0.05 0.2167 0.3833]
MF2='MediumLow':'itritype2',[0.1167 0.2833 0.45 0.2167 0.3833 0.55]
MF3='Medium':'itritype2',[0.2833 0.45 0.6167 0.3833 0.55 0.7167]
MF4='MediumHigh':'itritype2',[0.45 0.6167 0.7833 0.55 0.7167 0.8833]
MF5='High':'itritype2',[0.6167 0.7833 0.95 0.7167 0.8833 1.05]

[Rules]
1 1, 1 5 (1) : 1
1 2, 2 4 (1) : 1
1 3, 3 3 (1) : 1
2 1, 2 4 (1) : 1
2 2, 3 3 (1) : 1
2 3, 4 2 (1) : 1
3 1, 3 3 (1) : 1
3 2, 4 2 (1) : 1
3 3, 5 1 (1) : 1
```
End of file: IT2IterationDiversityAlphaRho.fis

Main Fuzzy Ant Colony Optimization with parameter adaptation through an interval type-2 fuzzy system applied to optimization of membership functions from fuzzy systems
Beginning of file: FACOT2forMF.m

```
%% Fuzzy Ant Colony Optimization for memership function optimization
function [bs_tour,bs_tlen,e_time]=FACOT2forMF()
    % Parameters
    maxit=100; % Maximum iterations
    m=0; % Number of ants
    alpha=1; beta=2; w=6;
    tic
    Q = 1;
    [n,m,dist,choice_info,pheromone,ant,base_params,base_fis] =
InitializeDataforMFO(m,alpha,beta);
    bsf_ant =
struct('tour_length',Inf,'tour',zeros(1,n),'visited',zeros(1,n));
    bs_tour = [];
    lengths = zeros(maxit,3);
    branching_avg = zeros(maxit,1);
```

```
% Read the fuzzy system for parameter adaptation
fis = readfis('IT2IterationDiversityAlphaRho.fis');
bs_u = 0;
u = w;
bs_rho = 0.1;
for i = 1:maxit
    ant =
ConstructSolutionsforMFO(n,m,choice_info,ant,dist,base_params,base_fis);
    [bs_tour,lengths,branching_avg] =
UpdateStatistics(bsf_ant,bs_tour,lengths,ant,i,pheromone,n,branching_avg,di
st);
    [bi_tlen,index] = min([ant(:).tour_length]);
    if(bi_tlen < bsf_ant.tour_length)
        bsf_ant = ant(index);
        assignin('base','Best',ant(index).fis);
    end
    bs_tlen = bsf_ant.tour_length;
    Iter = i/maxit;
    Diver = Diversity(ant);
    out = evalifistype2([Iter;Diver],fis);
    alpha = out(1); rho = out(2);
    [pheromone,choice_info] =
ASRankPheromoneAdaptativeUpdate(w,ant,bsf_ant,n,rho,pheromone,dist,alpha,be
ta,Q,bs_u,u,bs_rho);
    end
    e_time = toc;
end


%% Function to initialize the data of the ants
function [n,m,dist,choice_info,pheromone,ant,base_params,base_fis] =
InitializeDataforMFO(m,alpha,beta)
    dist = xlsread('mat.xls');
    n = size(dist,1);
    for i = 1:n; dist(i,i)=-Inf; end;
    if (m==0); m=n; end;
    base_fis = readfis('BASE.fis');
    base_params = struct('span_1',0,'x_1',0,'y_1',0,'x_2',0,'y_2',0,...
        'span_2',0,'span_3',0,'x_3',0,'y_3',0,'x_4',0,'y_4',0,...
        'span_4',0,'center_1',0,'span_5',0,'span_6',0,'center_2',0,...
'span_7',0,'center_3',0,'span_8',0,'span_9',0,'center_4',0,'span_10',0);
    ant(m) = struct('tour_length',0,'tour',zeros(1,n),'visited',...
        zeros(1,n),'fis',base_fis,'params',base_params);
```

```
    for i = 1:m
        ant(i) = struct('tour_length',0,'tour',zeros(1,n),...
            'visited',zeros(1,n),'fis',base_fis,'params',base_params);
    end
    pheromone = 0.1.*ones(n,n);
    for i = 1:n
        pheromone(i,i) = 0;
    end
    choice_info = ComputeChoiceInformation(n,pheromone,dist,alpha,beta);
end


%% Function to compute the choice information
function choice_info =
ComputeChoiceInformation(n,pheromone,dist,alpha,beta)
    choice_info = zeros(n,n);
    for i = 1:n
        for j = 1:i
            choice_info(i,j) =
(pheromone(i,j)^alpha)*((1/(dist(i,j)+0.1))^beta);
            choice_info(j,i) = choice_info(i,j);
        end
    end
end


%% Function to construct solutions for the ants
function ant =
ConstructSolutionsforMFO(n,m,choice_info,ant,dist,base_params,base_fis)
    for k = 1: m
        ant(k).visited = zeros(1,n);
        ant(k).params = base_params;
        ant(k).fis = base_fis;
    end
    step = 1;
    for k = 1:m
        r = randi(n,1);
        ant(k).tour(step) = r;
        ant(k).visited(r) = 1;
    end
    while step < n
        step = step + 1;
        for k = 1:m
            ant(k) = ASDecisionRule(n,ant(k),choice_info,step,dist);
```

```
            end
        end
        for k = 1:m
            ant(k).tour(n+1) = ant(k).tour(1);
            ant(k).tour_length = ComputeTourLength(ant(k));
        end
end


%% Fucntion for the AS decision rule
function ant = ASDecisionRule(n,ant,choice_info,i,dist)
        c = ant.tour(i-1);
        sum_probabilities = 0.0;
        selection_probability = zeros(1,n);
        for j = 1:n
            if ant.visited(j)
                selection_probability(j) = 0.0;
            else
                selection_probability(j) = choice_info(c,j);
                sum_probabilities = sum_probabilities +
selection_probability(j);
            end
        end
        r = sum_probabilities*rand(1);
        j = 1;
        p = selection_probability(j);
        while p < r
            j = j + 1;
            p = p + selection_probability(j);
        end
        ant.tour(i) = j;
        ant.visited(j) = 1;
        [ant.params,ant.fis] = UpdateFis(ant.params,ant.fis,dist,c,j);
end


%% Fitness function to calculate the length of the tour
function length = ComputeTourLength(ant)
        assignin('base','fis',ant.fis);
        simOut = sim('PRobot2');
        v_m_se = mean((data(:,2) - data(:,1)).^2);
        w_m_se = mean((data(:,4) - data(:,3)).^2);
        m_se = v_m_se + w_m_se;
        length = m_se;
end
```

```
%% Function to update the parameters of the ants to a fis
function [params,fis] = UpdateFis(params,fis,arc_val,i,j)
    switch i
        case 1
            params.span_1 = params.span_1 + arc_val(i,j);
            span_1 = params.span_1;
            x_1 = params.x_1;
            y_1 = params.y_1;
            fis = UpdateLSIMF(fis,span_1,x_1,y_1,1);
        case 2
            params.span_1 = params.span_1 + arc_val(i,j);
            span_1 = params.span_1;
            x_1 = params.x_1;
            y_1 = params.y_1;
            fis = UpdateLSIMF(fis,span_1,x_1,y_1,1);
        case 3
            params.x_1 = params.x_1 + arc_val(i,j);
            span_1 = params.span_1;
            x_1 = params.x_1;
            y_1 = params.y_1;
            fis = UpdateLSIMF(fis,span_1,x_1,y_1,1);
        case 4
            params.x_1 = params.x_1 + arc_val(i,j);
            span_1 = params.span_1;
            x_1 = params.x_1;
            y_1 = params.y_1;
            fis = UpdateLSIMF(fis,span_1,x_1,y_1,1);
        case 5
            params.y_1 = params.y_1 + arc_val(i,j);
            span_1 = params.span_1;
            x_1 = params.x_1;
            y_1 = params.y_1;
            fis = UpdateLSIMF(fis,span_1,x_1,y_1,1);
        case 6
            params.y_1 = params.y_1 + arc_val(i,j);
            span_1 = params.span_1;
            x_1 = params.x_1;
            y_1 = params.y_1;
            fis = UpdateLSIMF(fis,span_1,x_1,y_1,1);
        case 7
            params.x_2 = params.x_2 + arc_val(i,j);
            span_2 = params.span_2;
            x_2 = params.x_2;
            y_2 = params.y_2;
            fis = UpdateRSIMF(fis,span_2,x_2,y_2,1);
```

```
case 8
    params.x_2 = params.x_2 + arc_val(i,j);
    span_2 = params.span_2;
    x_2 = params.x_2;
    y_2 = params.y_2;
    fis = UpdateRSIMF(fis,span_2,x_2,y_2,1);
case 9
    params.y_2 = params.y_2 + arc_val(i,j);
    span_2 = params.span_2;
    x_2 = params.x_2;
    y_2 = params.y_2;
    fis = UpdateRSIMF(fis,span_2,x_2,y_2,1);
case 10
    params.y_2 = params.y_2 + arc_val(i,j);
    span_2 = params.span_2;
    x_2 = params.x_2;
    y_2 = params.y_2;
    fis = UpdateRSIMF(fis,span_2,x_2,y_2,1);
case 11
    params.span_2 = params.span_2 + arc_val(i,j);
    span_2 = params.span_2;
    x_2 = params.x_2;
    y_2 = params.y_2;
    fis = UpdateRSIMF(fis,span_2,x_2,y_2,1);
case 12
    params.span_2 = params.span_2 + arc_val(i,j);
    span_2 = params.span_2;
    x_2 = params.x_2;
    y_2 = params.y_2;
    fis = UpdateRSIMF(fis,span_2,x_2,y_2,1);
case 13
    params.span_3 = params.span_3 + arc_val(i,j);
    span_3 = params.span_3;
    x_3 = params.x_3;
    y_3 = params.y_3;
    fis = UpdateLSIMF(fis,span_3,x_3,y_3,2);
case 14
    params.span_3 = params.span_3 + arc_val(i,j);
    span_3 = params.span_3;
    x_3 = params.x_3;
    y_3 = params.y_3;
    fis = UpdateLSIMF(fis,span_3,x_3,y_3,2);
case 15
    params.x_3 = params.x_3 + arc_val(i,j);
    span_3 = params.span_3;
```

```
        x_3 = params.x_3;
        y_3 = params.y_3;
        fis = UpdateLSIMF(fis,span_3,x_3,y_3,2);
case 16
        params.x_3 = params.x_3 + arc_val(i,j);
        span_3 = params.span_3;
        x_3 = params.x_3;
        y_3 = params.y_3;
        fis = UpdateLSIMF(fis,span_3,x_3,y_3,2);
case 17
        params.y_3 = params.y_3 + arc_val(i,j);
        span_3 = params.span_3;
        x_3 = params.x_3;
        y_3 = params.y_3;
        fis = UpdateLSIMF(fis,span_3,x_3,y_3,2);
case 18
        params.y_3 = params.y_3 + arc_val(i,j);
        span_3 = params.span_3;
        x_3 = params.x_3;
        y_3 = params.y_3;
        fis = UpdateLSIMF(fis,span_3,x_3,y_3,2);
case 19
        params.x_4 = params.x_4 + arc_val(i,j);
        span_4 = params.span_4;
        x_4 = params.x_4;
        y_4 = params.y_4;
        fis = UpdateRSIMF(fis,span_4,x_4,y_4,2);
case 20
        params.x_4 = params.x_4 + arc_val(i,j);
        span_4 = params.span_4;
        x_4 = params.x_4;
        y_4 = params.y_4;
        fis = UpdateRSIMF(fis,span_4,x_4,y_4,2);
case 21
        params.y_4 = params.y_4 + arc_val(i,j);
        span_4 = params.span_4;
        x_4 = params.x_4;
        y_4 = params.y_4;
        fis = UpdateRSIMF(fis,span_4,x_4,y_4,2);
case 22
        params.y_4 = params.y_4 + arc_val(i,j);
        span_4 = params.span_4;
        x_4 = params.x_4;
        y_4 = params.y_4;
        fis = UpdateRSIMF(fis,span_4,x_4,y_4,2);
```

```
case 23
    params.span_4 = params.span_4 + arc_val(i,j);
    span_4 = params.span_4;
    x_4 = params.x_4;
    y_4 = params.y_4;
    fis = UpdateRSIMF(fis,span_4,x_4,y_4,2);
case 24
    params.span_4 = params.span_4 + arc_val(i,j);
    span_4 = params.span_4;
    x_4 = params.x_4;
    y_4 = params.y_4;
    fis = UpdateRSTMF(fis,span_4,x_4,y_4,2);
case 25
    params.center_1 = params.center_1 + arc_val(i,j);
    center_1 = params.center_1;
    span_5 = params.span_5;
    span_6 = params.span_6;
    fis = UpdateLSOMF(fis,center_1,span_5,span_6,1);
case 26
    params.center_1 = params.center_1 + arc_val(i,j);
    center_1 = params.center_1;
    span_5 = params.span_5;
    span_6 = params.span_6;
    fis = UpdateLSOMF(fis,center_1,span_5,span_6,1);
case 27
    params.span_5 = params.span_5 + arc_val(i,j);
    center_1 = params.center_1;
    span_5 = params.span_5;
    span_6 = params.span_6;
    fis = UpdateLSOMF(fis,center_1,span_5,span_6,1);
case 28
    params.span_5 = params.span_5 + arc_val(i,j);
    center_1 = params.center_1;
    span_5 = params.span_5;
    span_6 = params.span_6;
    fis = UpdateLSOMF(fis,center_1,span_5,span_6,1);
case 29
    params.span_6 = params.span_6 + arc_val(i,j);
    center_1 = params.center_1;
    span_5 = params.span_5;
    span_6 = params.span_6;
    fis = UpdateLSOMF(fis,center_1,span_5,span_6,1);
case 30
    params.center_2 = params.center_2 + arc_val(i,j);
    center_2 = params.center_2;
```

```
            span_6 = params.span_6;
            span_7 = params.span_7;
            fis = UpdateRSOMF(fis,center_2,span_6,span_7,1);
      case 31
            params.center_2 = params.center_2 + arc_val(i,j);
            center_2 = params.center_2;
            span_6 = params.span_6;
            span_7 = params.span_7;
            fis = UpdateRSOMF(fis,center_2,span_6,span_7,1);
      case 32
            params.span_7 = params.span_7 + arc_val(i,j);
            center_2 = params.center_2;
            span_6 = params.span_6;
            span_7 = params.span_7;
            fis = UpdateRSOMF(fis,center_2,span_6,span_7,1);
      case 33
            params.span_7 = params.span_7 + arc_val(i,j);
            center_2 = params.center_2;
            span_6 = params.span_6;
            span_7 = params.span_7;
            fis = UpdateRSOMF(fis,center_2,span_6,span_7,1);
      case 34
            params.center_3 = params.center_3 + arc_val(i,j);
            center_3 = params.center_3;
            span_8 = params.span_8;
            span_9 = params.span_9;
            fis = UpdateLSOMF(fis,center_3,span_8,span_9,2);
      case 35
            params.center_3 = params.center_3 + arc_val(i,j);
            center_3 = params.center_3;
            span_8 = params.span_8;
            span_9 = params.span_9;
            fis = UpdateLSOMF(fis,center_3,span_8,span_9,2);
      case 36
            params.span_8 = params.span_8 + arc_val(i,j);
            center_3 = params.center_3;
            span_8 = params.span_8;
            span_9 = params.span_9;
            fis = UpdateLSOMF(fis,center_3,span_8,span_9,2);
      case 37
            params.span_8 = params.span_8 + arc_val(i,j);
            center_3 = params.center_3;
            span_8 = params.span_8;
            span_9 = params.span_9;
            fis = UpdateLSOMF(fis,center_3,span_8,span_9,2);
```

```
        case 38
            params.span_9 = params.span_9 + arc_val(i,j);
            center_3 = params.center_3;
            span_8 = params.span_8;
            span_9 = params.span_9;
            fis = UpdateLSOMF(fis,center_3,span_8,span_9,2);
        case 39
            params.center_4 = params.center_4 + arc_val(i,j);
            center_4 = params.center_4;
            span_9 = params.span_9;
            span_10 = params.span_10;
            fis = UpdateRSOMF(fis,center_4,span_9,span_10,2);
        case 40
            params.center_4 = params.center_4 + arc_val(i,j);
            center_4 = params.center_4;
            span_9 = params.span_9;
            span_10 = params.span_10;
            fis = UpdateRSOMF(fis,center_4,span_9,span_10,2);
        case 41
            params.span_10 = params.span_10 + arc_val(i,j);
            center_4 = params.center_4;
            span_9 = params.span_9;
            span_10 = params.span_10;
            fis = UpdateRSOMF(fis,center_4,span_9,span_10,2);
        case 42
            params.span_10 = params.span_10 + arc_val(i,j);
            center_4 = params.center_4;
            span_9 = params.span_9;
            span_10 = params.span_10;
            fis = UpdateRSOMF(fis,center_4,span_9,span_10,2);
    end
end


%% Function to update the parameters in the left MF of the inputs
function fis = UpdateLSIMF(fis,span_1,x_1,y_1,input)
    param = -1 + ((0.475*(span_1/2))+0.475);
    span = 0 - param;
    m = span/2;
    c = span - m;
    x = param + ((m*(x_1/2))+c);
    y = (0.5*(y_1/2))+0.5;
    m = (y - 1)/((x - param)+eps);
    c = 1 - (m*param);
```

```
        fis.input(input).mf(1).params(3) = param;
        fis.input(input).mf(1).params(4) = (0 - c)/(m+eps);
        m = (1 - y)/((0 - x)+eps);
        c = 1 - (m*0);
        fis.input(input).mf(2).params(1) = (0 - c)/(m+eps);
end


%% Function to update the parameters in the left MF of the outputs
function fis = UpdateLSOMF(fis,center_1,span_5,span_6,output)
        fis.output(output).mf(1).params(2) = (0.5*(center_1/2))-0.5;
        fis.output(output).mf(1).params(3) =fis.output(output).mf(1).params(2)
+ (((0.475*(span_5/2))+0.525)/2);
        fis.output(output).mf(1).params(1) =fis.output(output).mf(1).params(2)
- (((0.475*(span_5/2))+0.525)/2);
        fis.output(output).mf(2).params(3) =fis.output(output).mf(2).params(2)
+ (((0.475*(span_6))+0.525)/2);
        fis.output(output).mf(2).params(1) =fis.output(output).mf(2).params(2)
- (((0.475*(span_6))+0.525)/2);
end


%% Function to update the parameters in the right MF of the inputs
function fis = UpdateRSIMF(fis,span_2,x_2,y_2,input)
        param = 1 - ((0.475*(span_2/2))+0.475);
        span = param;
        m = span/2;
        c = span - m;
        x = param - ((m*(x_2/2))+c);
        y = (0.5*(y_2/2))+0.5;
        m = (y - 1)/((x - 0)+eps);
        c = 1 - (m*0);
        fis.input(input).mf(2).params(3) = (0 - c)/m;
        fis.input(input).mf(3).params(2) = param;
        m = (1 - y)/((param - x)+eps);
        c = 1 - (m*param);
        fis.input(input).mf(3).params(1) = (0 - c)/m;
end


%% Function to update the parameters in the right MF of the outputs
function fis = UpdateRSOMF(fis,center_2,span_6,span_7,output)
        fis.output(output).mf(2).params(3) =fis.output(output).mf(2).params(2)
+ (((0.475.*(span_6))+0.525)/2);
        fis.output(output).mf(2).params(1) =fis.output(output).mf(2).params(2)
```

```
- (((0.475.*(span_6))+0.525)/2);
    fis.output(output).mf(3).params(2) = (0.5*(center_2/2))+0.5;
    fis.output(output).mf(3).params(3) = fis.output(output).mf(3).params(2)
+ (((0.475.*(span_7/2))+0.525)/2);
    fis.output(output).mf(3).params(1) = fis.output(output).mf(3).params(2)
- (((0.475.*(span_7/2))+0.525)/2);
end


%% Function to update the statistics
function [bs_tour,lengths,branching_avg] =
UpdateStatistics(bsf_ant,bs_tour,lengths,ant,i,pheromone,n,branching_avg,di
st)
        [bi_tlen,index] = min([ant(:).tour_length]);
        bs_tlen = bsf_ant.tour_length;
        if(bi_tlen < bs_tlen)
            bs_tlen = bi_tlen;
            bs_tour = ant(index).tour;
        end
        m_tlen = mean([ant(:).tour_length]);
        lengths(i,:) = [m_tlen bi_tlen bs_tlen];
        branching = node_branching(0.05,pheromone,n,dist);
        branching_avg(i,1) = branching;
end


%% Function calculare the node branching
function branching = node_branching(l,pheromone,n,dist)
    [~,nn_list] = sort(dist,2,'ascend');
    nn_length = n;
    num_branches = zeros(n,1);
    for m = 1:n
        minimum = pheromone(m,nn_list(m,2));
        maximum =  pheromone(m,nn_list(m,2));
        for i = 2:nn_length
            if pheromone(m,nn_list(m,i)) > maximum
                maximum = pheromone(m,nn_list(m,i));
            end
            if pheromone(m,nn_list(m,i)) < minimum
                minimum = pheromone(m,nn_list(m,i));
            end
        end
        cutoff = minimum + 1 * (maximum - minimum);
        for i = 1:nn_length
            if pheromone(m,nn_list(m,i)) > cutoff
```

```
                    num_branches(m) = num_branches(m) + 1;
             end
        end
    end
    branching = mean(num_branches);
end


%% Function to update the pheromone adaptovely
function [pheromone,choice_info] =
ASRankPheromoneAdaptativeUpdate(w,ant,bsf_ant,n,rho,pheromone,dist,alpha,be
ta,Q,bs_u,u,bs_rho)
    rho_mat = GetRhoMat(bsf_ant,n,rho,bs_rho);
    pheromone = AdaptativeEvaporation(n,rho_mat,pheromone);
    pheromone =
ASRankBSAntAdaptativePheromoneDeposit(n,bsf_ant,pheromone,Q,bs_u);
    [V,I] = sort([ant(:).tour_length]);
    U = V;
    if(ismember(bsf_ant.tour_length,U(1:w-1)))
       U = U(2:length(U));
    end
    L = zeros(w-1,1);
    for k = 1:w-1
       L(k) = find(V == U(k),1);
       pheromone =
ASRankAdaptativePheromoneDeposit(w,k,n,ant(I(L(k))),pheromone,Q,u);
    end
    choice_info = ComputeChoiceInformation(n,pheromone,dist,alpha,beta);
end


%% Function to calculate the rho matrix
function rho_mat = GetRhoMat(bsf_ant,n,rho,b_rho)
    b_arcs = [];
    for y = 1:n
       b_arcs = [b_arcs; [bsf_ant.tour(y) bsf_ant.tour(y+1)]];
    end
    rho_mat = ones(n,n).*rho;
    for x = 1:size(b_arcs,1)
       rho_mat(b_arcs(x,1),b_arcs(x,2)) = b_rho;
```

```
            rho_mat(b_arcs(x,2),b_arcs(x,1)) =
rho_mat(b_arcs(x,1),b_arcs(x,2));
    end
end


%% Function to evaporate the pheromone of the arcs
function pheromone = AdaptativeEvaporation(n,rho_mat,pheromone)
    for i = 1:n
        for j = 1:i
            pheromone(i,j) = (1-rho_mat(i,j))*pheromone(i,j);
            pheromone(j,i) = pheromone(i,j);
        end
    end
end


%% Function to deposite pheromone of the best ant
function pheromone =
ASRankBSAntAdaptativePheromoneDeposit(n,bsf_ant,pheromone,Q,bs_u)
    phe_inc = bs_u*(Q/bsf_ant.tour_length);
    for i = 1:n
        j = bsf_ant.tour(i);
        l = bsf_ant.tour(i + 1);
        pheromone(j,l) = pheromone(j,l) + phe_inc;
        pheromone(l,j) = pheromone(j,l);
    end
end


%% Function to deposit pheromone of all ants
function pheromone =
ASRankAdaptativePheromoneDeposit(w,r,n,ant,pheromone,Q,u)
    phe_inc = ((w-r)/(w-1))*(u)*(Q/ant.tour_length);
    for i = 1:n
        j = ant.tour(i);
        l = ant.tour(i + 1);
        pheromone(j,l) = pheromone(j,l) + phe_inc;
        pheromone(l,j) = pheromone(j,l);
    end
end
```

End of file: FACOT2forMF.m

Fuzzy inference system used as basis controller for the autonomous mobile robot problem

Beginning of file: _BASE.fis_

```
[System]
Name='BASE'
Type='mamdani'
Version=2.0
NumInputs=2
NumOutputs=2
NumRules=9
AndMethod='min'
OrMethod='max'
ImpMethod='min'
AggMethod='max'
DefuzzMethod='centroid'

[Input1]
Name='ev'
Range=[-1 1]
NumMFs=3
MF1='N':'trapmf',[-1 -1 -1 -1]
MF2='Z':'trimf',[-1 0 1]
MF3='P':'trapmf',[1 1 1 1]

[Input2]
Name='ew'
Range=[-1 1]
NumMFs=3
MF1='N':'trapmf',[-1 -1 -1 -1]
MF2='Z':'trimf',[-1 0 1]
MF3='P':'trapmf',[1 1 1 1]

[Output1]
Name='T1'
Range=[-1 1]
NumMFs=3
MF1='N':'trimf',[-1 -1 -1]
MF2='Z':'trimf',[0 0 0]
MF3='P':'trimf',[1 1 1]

[Output2]
Name='T2'
Range=[-1 1]
```

```
NumMFs=3
MF1='N':'trimf',[-1 -1 -1]
MF2='Z':'trimf',[0 0 0]
MF3='P':'trimf',[1 1 1]

[Rules]
1 1, 1 1 (1) : 1
1 2, 1 2 (1) : 1
1 3, 1 3 (1) : 1
2 1, 2 1 (1) : 1
2 2, 2 2 (1) : 1
2 3, 2 3 (1) : 1
3 1, 3 1 (1) : 1
3 2, 3 2 (1) : 1
3 3, 3 3 (1) : 1
```

End of file: BASE.fis

Main Fuzzy Gravitational Search Algorithm with parameter adaptation through an interval type-2 fuzzy system applied to the minimization of benchmark mathematical functions

Beginning of file: FGSAT2forBMF.m

```matlab
%% Fuzzy Gravitational Search Algorithm for benchmark mathematical
functions
function [Fbest,Lbest,BestChart,MeanChart]=FGSAT2forBMF()
    F_index = 1; % functions 1 to 15
    N=50; % Number of agents
    max_it=1000; % Maximum iterations
    min_flag=1;
    Rpower=1;
    Rnorm=2;
    [dim,low,up] = TestParam(F_index);
    X=initialization(dim,N,up,low);
    BestChart=[]; MeanChart=[];
    V=zeros(N,dim);
    % Read the fuzzy system for parameter adaptation
    fis=readfis('IT2IterationDiversityAlphaKbest.fis');
    for iteration=1:max_it
        X=space_bound(X,up,low); % Checking allowable range.
        fitness=evaluateF(X,F_index);
        if min_flag==1
            [best best_X]=min(fitness); %minimization.
        else
            [best best_X]=max(fitness); %maximization.
        end
```

```
        if iteration==1; Fbest=best;Lbest=X(best_X,:); end;
        if min_flag==1
            if best<Fbest; Fbest=best;Lbest=X(best_X,:); end;
        else
            if best>Fbest; Fbest=best;Lbest=X(best_X,:); end;
        end
        [M]=massCalculation(fitness,min_flag);
        It=iteration/max_it;
        Diver = Diversity(X,best_X);
        out=evalifistype2([It, Diver],fis);
        alfa=out(1); kbest=out(2);
        G=Gconstant(iteration,max_it,alfa);
        a=Gfield(M,X,G,Rnorm,Rpower,kbest);
        [X,V]=move(X,a,V);
    end %End iterations
end


%% This function initializes the position of the agents in the search
space, randomly
function [X]=initialization(dim,N,up,down)
    if size(up,2)==1; X=rand(N,dim).*(up-down)+down; end;
    if size(up,2)>1
        for i=1:dim
            high=up(i);low=down(i);
            X(:,i)=rand(N,1).*(high-low)+low;
        end
    end
end


%% This function checks the search space boundaries for agents
function  X=space_bound(X,up,low)
    [N,dim]=size(X);
    for i=1:N
Tp=X(i,:)>up;Tm=X(i,:)<low;X(i,:)=(X(i,:).*(~(Tp+Tm)))+((rand(1,dim).*(up-
low)+low).*(Tp+Tm));
    end
end


%% This function Evaluates the agents
function fitness=evaluateF(X,F_index)
    [N,dim]=size(X);
    fitness = zeros(1,N);
    for i=1:N
        L=X(i,:);
        fitness(i)=TestFunctions(L,F_index);
    end
end
```

```
%% This function calculates the mass of each agent
function [M]=massCalculation(fit,min_flag)
    Fmax=max(fit); Fmin=min(fit); %Fmean=mean(fit);
    N = size(fit,2);
    if Fmax==Fmin
      M=ones(N,1);
    else
      if min_flag==1 %for minimization
        best=Fmin;worst=Fmax;
      else %for maximization
        best=Fmax;worst=Fmin;
      end
      M=(fit-worst)./(best-worst);
    end
    M=M./sum(M);
end


%% This function calculates Gravitational constant
function G=Gconstant(iteration,max_it,alfa)
    G0=100;
    G=G0*exp(-alfa*iteration/max_it);
end


%% This function calculates the acceleration of each agent in gravitational
field
function a=Gfield(M,X,G,Rnorm,Rpower,kbest)
    [N,dim]=size(X);
    kbest=round(N*kbest/100);
    [~, ds]=sort(M,'descend');
    for i=1:N
        E(i,:)=zeros(1,dim);
        for ii=1:kbest
            j=ds(ii);
            if j~=i
                R=norm(X(i,:)-X(j,:),Rnorm); %Euclidian distanse.
                for k=1:dim
                    E(i,k)=E(i,k)+rand*(M(j))*((X(j,k)-
X(i,k))/(R^Rpower+eps));
                end
            end
        end
    end
```

```
        a=E.*G; %note that Mp(i)/Mi(i)=1
end


%% This function updates the velocity and position of agents
function [X,V]=move(X,a,V)
        [N,dim]=size(X);
        V=rand(N,dim).*V+a;
        X=X+V;
end
```

End of file: FGSAT2forBMF.m

Interval type-2 fuzzy system for parameter adaptation in GSA
Beginning of file: IT2IterationDiversityAlphaKbest.fis

```
[System]
Name='IT2IterationDiversityAlphaKbest'
Type='mamdani'
Version=2.0
NumInputs=2
NumOutputs=2
NumRules=9
AndMethod='min'
OrMethod='max'
ImpMethod='min'
AggMethod='max'
DefuzzMethod='centroid'

[Input1]
Name='Iteration'
Range=[0 1]
NumMFs=3
MF1='Low':'itritype2',[-0.62 -0.12 0.38 -0.38 0.12 0.62]
MF2='Medium':'itritype2',[-0.12 0.38 0.88 0.12 0.62 1.12]
MF3='High':'itritype2',[0.38 0.88 1.38 0.62 1.12 1.62]

[Input2]
Name='Diversity'
Range=[0 1]
NumMFs=3
MF1='Low':'itritype2',[-0.62 -0.12 0.38 -0.38 0.12 0.62]
MF2='Medium':'itritype2',[-0.12 0.38 0.88 0.12 0.62 1.12]
MF3='High':'itritype2',[0.38 0.88 1.38 0.62 1.12 1.62]
```

```
[Output1]
Name='Alfa'
Range=[0 100]
NumMFs=5
MF1='Low':'itritype2',[-30 -5 20 -20 5 30]
MF2='MediumLow':'itritype2',[-5 20 45 5 30 55]
MF3='Medium':'itritype2',[20 45 70 30 55 80]
MF4='MediumHigh':'itritype2',[45 70 95 55 80 105]
MF5='High':'itritype2',[70 95 120 80 105 130]

[Output2]
Name='kbest'
Range=[0 1]
NumMFs=5
MF1='Low':'itritype2',[-0.29 -0.04 0.21 -0.21 0.04 0.29]
MF2='MediumLow':'itritype2',[-0.04 0.21 0.46 0.04 0.29 0.54]
MF3='Medium':'itritype2',[0.21 0.46 0.71 0.29 0.54 0.79]
MF4='MediumHigh':'itritype2',[0.46 0.71 0.96 0.54 0.79 1.04]
MF5='High':'itritype2',[0.71 0.96 1.21 0.79 1.04 1.29]

[Rules]
1 1, 3 5 (1) : 1
1 2, 2 4 (1) : 1
1 3, 1 3 (1) : 1
2 1, 4 4 (1) : 1
2 2, 3 3 (1) : 1
2 3, 2 2 (1) : 1
3 1, 5 3 (1) : 1
3 2, 4 2 (1) : 1
3 3, 3 1 (1) : 1
```
End of file: IT2IterationDiversityAlphaKbest.fis

Main Fuzzy Gravitational Search Algorithm with parameter adaptation through an interval type-2 fuzzy system applied to optimization of membership function from fuzzy systems
Beginning of file: FGSAT2forMF.m

```
%% Fuzzy Gravitational Search Algorithm for optimization of FIS
function [Fbest,Lbest,BestChart,MeanChart] = GSA()
    N = 30; % N: Number of agents.
    max_it = 100; % max_it: Maximum number of iterations (T).
    Rnorm = 2; %Rnorm: Norm in eq.8.
    Rpower = 1; % Rpower: power of 'R' in eq.7.
    min_flag=1; % 1: minimization, 0: maximization
    ControlFis=readfis('./Shower.fis');
```

```
    dim = nPar(ControlFis);
    Lim = Limits(ControlFis,dim);
    X = initialization(dim,N,Lim);
    BestChart=[]; MeanChart=[];
    V=zeros(N,dim);
    % Read the fuzzy system for parameter adaptation
    fis=readfis('./IT2IterDiverAlphaKbest.fis');
    for iteration=1:max_it
        X = ValidatePopulationGSA(ControlFis,X,Lim);
        fitness = FitnessFunctionGSA(ControlFis,X);
        disp(num2str(iteration))
        if min_flag==1
            [best best_X]=min(fitness); %minimization.
        else
            [best best_X]=max(fitness); %maximization.
        end
        if iteration==1; Fbest=best;Lbest=X(best_X,:); end;
        if min_flag==1
            if best<Fbest; Fbest=best;Lbest=X(best_X,:); end;
        else
            if best>Fbest; Fbest=best;Lbest=X(best_X,:); end;
        end
        [M]=massCalculation(fitness,min_flag);
        It = iteration/max_it;
        Diver = Diversity(X,best_X);
        out = evalifistype2([It,Diver],fis);
        alfa = out(1); kbest = out(2);
        G=Gconstant(iteration,max_it,alfa);
        a=Gfield(M,X,G,Rnorm,Rpower,kbest);
        [X,V]=move(X,a,V);
    end %End iterations
end


%% This function initializes the position of the agents in the search
space, randomly.
function [X] = initialization(dim,N,Lim)
    c = 1;
    X = zeros(N,dim);
    for i=1:dim
        high = Lim(c); low = Lim(c+1);
        X(:,i) = rand(N,1).*(high-low)+low;
        c = c+2;
    end
end
```

```
%% This function calculates the mass of each agent
function [M]=massCalculation(fit,min_flag)
    Fmax=max(fit); Fmin=min(fit);
    N = size(fit,2);
    if Fmax==Fmin
      M=ones(N,1);
    else
      if min_flag==1 %for minimization
        best=Fmin;worst=Fmax;
      else %for maximization
        best=Fmax;worst=Fmin;
      end
      M=(fit-worst)./(best-worst);
    end
    M=M./sum(M);
end


%% This function calculates Gravitational constant
function G=Gconstant(iteration,max_it,alfa)
    G0=100;
    G=G0*exp(-alfa*iteration/max_it);
end


%% This function calculates the acceleration of each agent in gravitational
field
function a=Gfield(M,X,G,Rnorm,Rpower,kbest)
    [N,dim] = size(X);
    kbest = kbest*100;
    kbest=round(N*kbest/100);
    [~, ds]=sort(M,'descend');
    for i=1:N
        E(i,:)=zeros(1,dim);
        for ii=1:kbest
            j=ds(ii);
            if j~=i
                R=norm(X(i,:)-X(j,:),Rnorm); %Euclidian distanse.
                for k=1:dim
                    E(i,k)=E(i,k)+rand*(M(j))*((X(j,k)-
X(i,k))/(R^Rpower+eps));
                end
            end
```

```
            end
        end
        a=E.*G;
end


%% This function updates the velocity and position of agents
function [X,V]=move(X,a,V)
        [N,dim]=size(X);
        V=rand(N,dim).*V+a;
        X=X+V;
end
```

End of file: FGSAT2forMF.m

Function used as fitness function for GSA in the membership function optimization
Beginning of file: FitnessFunctionGSA.m

```
%% Fitness function for FIS optimization
% FIS = Basis fuzzy system
% Agents = Is the population of agents
function Fitness = FitnessFunctionGSA(FIS,Agents)
        [N,dim] = size(Agents);
        Fitness = zeros(1,N);
        for i=1:N
            fprintf '.'
            Agent = Agents(i,:);
            FIS = Agent2FIS(FIS,Agent);
            assignin('base','ShowerFIS',FIS);
            sim('ShowerPlant'); % Simulation
            TempMSE = mean((Temp(:,2) - Temp(:,3)).^2); % Temperature Error
            FlowMSE = mean((Flow(:,2) - Flow(:,3)).^2); % Flow Error
            Fitness(i) = TempMSE + FlowMSE;
        end
end
```

End of file: FitnessFunctionGSA.m

Function used to convert an agent into a fuzzy controller
Beginning of file: Agent2FIS.m

```
%% Convert an Agent to a FIS
% FIS = Basis fuzzy system
% Agent = Contains the parameters for the FIS
% The agent needs to be evaluated before
function FIS = Agent2FIS(FIS,Agent)
        cont = 1;
```

```
    for i=1:size(FIS.input,2) % For each input
        for j=1:size(FIS.input(i).mf,2) % For each MFs
            if strcmp(FIS.input(i).mf(j).type,'trimf') % Triangular
                FIS.input(i).mf(j).params = Agent(cont:cont+2);
                cont=cont+3;
            elseif strcmp(FIS.input(i).mf(j).type,'trapmf') % Trapezoidal
                FIS.input(i).mf(j).params = Agent(cont:cont+3);
                cont=cont+4;
            end
        end
    end
    for i=1:size(FIS.output,2) % For each output
        for j=1:size(FIS.output(i).mf,2) % For each MFs
            if strcmp(FIS.output(i).mf(j).type,'trimf') % Triangular
                FIS.output(i).mf(j).params = Agent(cont:cont+2);
                cont=cont+3;
            elseif strcmp(FIS.output(i).mf(j).type,'trapmf') % Trapezoidal
                FIS.output(i).mf(j).params = Agent(cont:cont+3);
                cont=cont+4;
            end
        end
    end
end
```

End of file: Agent2FIS.m

Function to calculate the number of parameters of the membership functions to be optimized with GSA
Beginning of file: nPar.m

```
%% Calculate the parameters of the MFs to be optimized
% FIS = Basis fuzzy system
function npar = nPar(FIS)
    cont=0;
    for i=1:size(FIS.input,2) % For each input
        for j=1:size(FIS.input(i).mf,2) % For each MFs
            if strcmp(FIS.input(i).mf(j).type,'trimf') % Triangular
                cont=cont+3;
            elseif strcmp(FIS.input(i).mf(j).type,'trapmf') % Trapezoidal
                cont=cont+4;
            end
        end
    end
```

```
    for i=1:size(FIS.output,2) % For each output
        for j=1:size(FIS.output(i).mf,2) % For each MFs
            if strcmp(FIS.output(i).mf(j).type,'trimf') % Triangular
                cont=cont+3;
            elseif strcmp(FIS.output(i).mf
(j).type,'trapmf') % Trapezoidal
                cont=cont+4;
            end
        end
    end
    npar = cont;
end
```

End of file: nPar.m

Function to search for the limits of each parameter of the membership functions
Beginning of file: Limits.m

```
%% Establishes the limits of parameters of the MFs
% FIS = Basis fuzzy system
% npar = Total number of parameters
function Lim = Limits(fis,npar)
    Lim = zeros(1,npar*2);
    x=1;
    for i=1:length(fis.input)
        for j=1:length(fis.input(i).mf)
            Lim(x) = fis.input(i).mf(j).params(1);
            Lim(x+1) = fis.input(i).mf(j).params(2);
            Lim(x+2) = fis.input(i).mf(j).params(2);
            Lim(x+3) = fis.input(i).mf(j).params(3);
            x = x+4;
        end
    end
    for i=1:length(fis.output)
        for j=1:length(fis.output(i).mf)
            Lim(x) = fis.output(i).mf(j).params(1);
            Lim(x+1) = fis.output(i).mf(j).params(2);
            Lim(x+2) = fis.output(i).mf(j).params(2);
            Lim(x+3) = fis.output(i).mf(j).params(3);
            x = x+4;
        end
```

```
        end
end
```

End of file: Limits.m

Function to validate the parameters contained in an agent
Beginning of file: ValidateAgent.m

```
%% Validate an agent
% FIS = Basis fuzzy system
% Agent = Contains the parameters to be evaluated
function Agent = ValidateAgent(FIS,Agent)
    cont=0;
    for i=1:size(FIS.input,2) % For each input
        for j=1:size(FIS.input(i).mf,2) % For each MFs
            if strcmp(FIS.input(i).mf(j).type,'trimf') % Triangular
                x=sort([Agent(cont+1),Agent(cont+2),Agent(cont+3)]);
                Agent(cont+1) = x(1);
                Agent(cont+2) = x(2);
                Agent(cont+3) = x(3);
                if cont>0
                    if Agent(cont)<Agent(cont+1)
                        A1=Agent(cont+1);
                        Agent(cont+1)=Agent(cont);
                        Agent(cont)=A1;
                    end
                end
                cont=cont+3;
            elseif strcmp(FIS.input(i).mf(j).type,'trapmf') % Trapezoidal
x=sort([Agent(cont+1),Agent(cont+2),Agent(cont+3),Agent(cont+4)]);
                Agent(cont+1) = x(1);
                Agent(cont+2) = x(2);
                Agent(cont+3) = x(3);
                Agent(cont+4) = x(4);
                if cont>0
                    if Agent(cont)<Agent(cont+1)
                        A1=Agent(cont+1);
                        Agent(cont+1)=Agent(cont);
                        Agent(cont)=A1;
                    end
                end
                cont=cont+4;
            end
        end
```

```
        end
    for i=1:size(FIS.output,2) % For each output
        out = [1,2,4,3,5,7,6,8,10,9,11,13,12,14,15];
        x = Agent(cont+1:cont+15);
        x = sort(x);
        for j=1:15
            z=out(j);
            Agent(cont+j) = x(z);
        end
        cont = cont+15;
    end
end
```

End of file: ValidateAgent.m

Function to validate if the parameters contained in the agents are between the limits of the search space

Beginning of file: ValidatePopulationGSA.m

```
%% Validate the population of GSA
% FIS = Basis fuzzy system
% Pop = Population of Agents
% Lim = Limits of each MF
function Pop = ValidatePopulationGSA(FIS,Pop,Lim)
    [x,y] = size(Pop);
    for i=1:x
      k=1;
      for j=1:y
        if Pop(i,j) < Lim(k) || Pop(i,j) > Lim(k+1)
            Pop(i,j) = rand(1)*(Lim(k+1)-Lim(k))+Lim(k);
        end
        k=k+2;
      end
      Pop(i,:) = ValidateAgent(FIS,Pop(i,:));
    end
end
```

End of file: ValidatePopulationGSA.m

Fuzzy system used as basis controller for the automatic temperature control in a shower

Beginning of file: shower.fis

```
% $Revision: 1.1 $
[System]
Name = 'shower'
Type = 'mamdani'
```

```
NumInputs = 2
NumOutputs = 2
NumRules = 9
AndMethod = 'min'
OrMethod = 'max'
ImpMethod = 'min'
AggMethod = 'max'
DefuzzMethod = 'centroid'

[Input1]
Name = 'temp'
Range = [-20 20]
NumMFs = 3
MF1='cold':'trapmf',[-30 -30 -15 0]
MF2='good':'trimf',[-10 0 10 0]
MF3='hot':'trapmf',[0 15 30 30]

[Input2]
Name = 'flow'
Range = [-1 1]
NumMFs = 3
MF1='soft':'trapmf',[-3 -3 -0.8 0]
MF2='good':'trimf',[-0.4 0 0.4 0]
MF3='hard':'trapmf',[0 0.8 3 3]

[Output1]
Name = 'cold'
Range = [-1 1]
NumMFs = 5
MF1='closeFast':'trimf',[-1 -0.6 -0.3]
MF2='closeSlow':'trimf',[-0.6 -0.3 0]
MF3='steady':'trimf',[-0.3 0 0.3]
MF4='openSlow':'trimf',[0 0.3 0.6]
MF5='openFast':'trimf',[0.3 0.6 1]

[Output2]
Name = 'hot'
Range = [-1 1]
NumMFs = 5
MF1='closeFast':'trimf',[-1 -0.6 -0.3]
MF2='closeSlow':'trimf',[-0.6 -0.3 0]
MF3='steady':'trimf',[-0.3 0 0.3]
MF4='openSlow':'trimf',[0 0.3 0.6]
```

```
MF5='openFast':'trimf',[0.3 0.6 1]
```

[Rules]
```
1 1, 4 5 (1) : 1
1 2, 2 4 (1) : 1
1 3, 1 2 (1) : 1
2 1, 4 4 (1) : 1
2 2, 3 3 (1) : 1
2 3, 2 2 (1) : 1
3 1, 5 4 (1) : 1
3 2, 4 2 (1) : 1
3 3, 2 1 (1) : 1
```
End of file: shower.fis

Index

© The Author(s) 2018
F. Olivas et al., *Dynamic Parameter Adaptation for Meta-Heuristic Optimization Algorithms Through Type-2 Fuzzy Logic*, SpringerBriefs in Computational Intelligence https://doi.org/10.1007/978-3-319-70851-5

>

Printed in the United States
By Bookmasters